2nd International Conference on Materials Science and Nanotechnology (ICMSNT 2017) and International Symposium on Material Science and Technology (ISMST 2017)

MATEC Web of Conferences Volume 109 (2017)

Auckland, New Zealand
19 - 22 April 2017

Editors:

Zhan Chen
Timotius Pasang
Kin Tak Alan Lau
Min-hao Zhu

ISBN: 978-1-5108-4116-1

Printed from e-media with permission by:

Curran Associates, Inc.
57 Morehouse Lane
Red Hook, NY 12571

Some format issues inherent in the e-media version may also appear in this print version.

Printed by Curran Associates, Inc. (2017)

For additional information, please contact EDP Sciences – Web of Conferences at the address below.

EDP Sciences – Web of Conferences
17, Avenue du Hoggar
Parc d'Activité de Courtabœuf
BP 112
F-91944 Les Ulis Cedex A
France

Phone: +33 (0) 1 69 18 75 75
Fax: +33 (0) 1 69 28 84 91

contact-edps@webofconferences.org

Additional copies of this publication are available from:

Curran Associates, Inc.
57 Morehouse Lane
Red Hook, NY 12571 USA
Phone: 845-758-0400
Fax: 845-758-2633
Email: curran@proceedings.com
Web: www.proceedings.com

TABLE OF CONTENTS

ISSN (electronic edition): 2261-236X

Preface

The 2nd International Conference on Materials Science and Nanotechnology (ICMSNT 2017) and 2017 2nd International Symposium on Material Science and Technology (ISMST 2017) were held at Auckland University of Technology, Auckland, New Zealand during April 19-21, 2017. The conference was organized by Hong Kong Society of Mechanical Engineers (HKSME). During the conference, delegates from India, China, Malaysia, Thailand, South Korea, Taiwan, Portugal, United Kingdom, New Zealand and other places gathered together and presented their research in a variety of ICMSNT and ISMST topics. The organizer (HKSME) would like to thank all of the presenters who have come far to NZ and have contributed to the enjoyable and valuable discussions during the conference.

Peer-reviewing of all submitted papers was conducted before the papers were selected for presentation in the conference. For each paper included in this proceeding, the paper was peer-reviewed by 2-3 experts from the conference technical committee following the normal refereeing procedure. The conference proceedings containing 27 papers are divided into 5 parts: Properties of Engineering Materials, Material/Mechanical Behaviours during Materials Processing, Materials Processing Technologies, Properties and Service Behaviours of Materials and Nanomaterials, Modelling to Predict Mechanical Behaviours and Other Technologies. The organizer (HKSME) would like to thank the contributions made by the technical committee. They gave their times and reviewed all the papers in their respective fields of expertise.

Prof. Zhan Chen

Auckland University of Technology, NZ

ICMSNT 2017 Committee

Conference Committee Co-Chairs

Prof. Chen Zhan, Auckland University of Technology, New Zealand
Prof. Timotius Pasang, Auckland University of Technology, New Zealand
Prof. LAU Kin Tak Alan, Swinburne University of Technology, Australia
Prof. Shinhoo Kang, Seoul National University, South Korea

Program Committee Chairs

Prof. Ramesh Singh, University of Malaya, Malaysia
Prof. Min-Hao Zhu, Southwest Jiaotong University, China

Publication Chairs

Prof. Song-Jeng Huang, National Taiwan University of Science and Technology, Taiwan
Prof. Osman Adiguzel, Firat University, Turkey

Local Chair

Dr. Andrew Hilton, Auckland University of Technology, New Zealand

International Technical Committees

Prof. John Kennedy, Principal Scientist, GNS Science & Honorary Professor, University of Waikato, New Zealand
Prof. Yongxiang Leng, Southwest Jiaotong University, China
Prof. Liangchi Zhang, University of New South Wales, Australia
Dr. Namasivayam Navaranjan, Universiti Teknologi Brunei, Brunei Darussalam
Dr. YANG Weifeng, Institute of Materials Research and Engineering, Singapore
Dr. Jun Yang, University of Western Ontario, Canada
Dr. Dong Zhili,Nanyang Technological University, Singapore
Dr. Rajesh Kumar Bhushan, A Central University, India
Dr. MohammadReza Khanzadeh, Islamic Azad University,Majlesi Branch, Iran
Dr.Thirumany Sritharan,Nanyang Technological University, Singapore
Dr. YIP MUM WAI, Tunku Abdul Rahman University College, Malaysia
Dr. Ruitao Lv, Tsinghua University,China
Dr. Teo Hang Tong Edwin, Nanyang Technological University, Singapore
Dr. Bo Ling, Nanometrics Inc. USA

edp sciences

MATEC Web of Conferences

Statement of Peer review

In submitting conference proceedings to MATEC Web of Conferences, I certify to the Publisher that I adhere to the **Policy on Publishing Integrity** of the journal in order to safeguard good scientific practice in publishing.

1. All articles have been subjected to peer review administered by the proceedings editors.
2. Reviews have been conducted by expert referees, who have been requested to provide unbiased and constructive comments aimed, whenever possible, at improving the work.
3. Proceedings editors have taken all reasonable steps to ensure the quality of the materials they publish and their decision to accept or reject a paper for publication has been based only on the merits of the work and the relevance to the journal.

Title, date and place of the conference

2017 2nd International Conference on Materials Science and Nanotechnology (ICMSNT 2017)
2017 2nd International Symposium on Material Science and Technology (ISMST 2017)

Auckland University of Technology, Auckland, New Zealand, April 19-22, 2017

Proceedings editor(s):

Prof. Zhan Chen Prof. Timotius Pasang Prof.LAU Kin Tak Alan

30 April 2017

Prof.ZHU Min-hao

Date and editor's signature April 30th, 2017

Optimization of Mechanical Properties of Silver Nanoparticles (AgNPS)-Loaded Chitosan/Polylactic Acid (PLA) Biofilms by Using Response Surface Methodology (RSM)

A. Shaari[1], M.N. Abu Seman[1,a] and C.K.M. Faizal[2]

[1] *Faculty of Chemical Engineering & Natural Resources, Universiti Malaysia Pahang (UMP), Lebuhraya Tun Razak, 26300 Gambang, Kuantan, Pahang, Malaysia.*
[2] *Faculty of Engineering Technology, Universiti Malaysia Pahang (UMP), Lebuhraya Tun Razak, 26300 Gambang, Kuantan, Pahang, Malaysia.*

Abstract. Fabrication of silver nanoparticles-loaded chitosan-polylactic acid based biofilms was successfully employed for investigating the optimal of mechanical properties (i.e. tensile strength and elongation at break) of biofilms using central composite design (CCD), response surface methodology (RSM). In this study, only two factors that influences the biofilm mechanical properties were selected namely concentration of polyethelene glycol 400 (PEG) and percentage volume of polylactic acid (PLA)/chitosan. Analysis of results was performed by using response surface methodology (RSM) to avoid the traditional one-factor-at-a-time experiments. Common statistical tools such as analysis of variance (ANOVA) and response surface plot were used to determine the optimal tensile strength and elongation at break responses. Central composite design (CCD) builds a response surface for mechanical properties of biofilms optimization. From the results of statistical analysis, it could be concluded that the optimal conditions for mechanical properties of biofilms were 7.93% w/w concentration of polyethylene glycol (PEG) and 28.79%/71.21% percentage volume of polylactic acid (PLA)/chitosan. At this optimum stage, 7.99 MPa of tensile strength and 32.6 % elongation at break were obtained. Then, results of verification process have shown that the percentage errors are 2.08% for tensile strength and 3.89% elongation at break, respectively.

1 Introduction

Recently, high demands and low cost of production are the reasons of synthetic polymer has become increasingly valuable [1] especially in food packaging and medical field. These synthetic polymers are not biodegradable; it is finally end up in landfill sites and produce very harmful gases that can cause environmental pollutions and soil contaminations. The burning of synthetic polymers can cause the release of dioxin which is highly toxic that will cause chronic disease (i.e: cancer) through inhalation. Therefore, it is very crucial to discover other biodegradable films to replace the use of synthetic polymer. The biodegradable films were not meant to totally substitute the synthetic polymers. Nevertheless, the using of biodegradable films as packaging materials is still one of the most hopeful

[a] Corresponding author: mazrul@ump.edu.my

approaches for effective methods of maintaining in food quality and medical application. By using the biodegradable films for food product packaging and medical field, the industry are facing with the problems of short shelf life period due to low antimicrobial resistance [2]. Due to that, the inclusion of the active compounds in the films becomes an innovative approach in order to avoid food product contamination by microbial [3]. Silver nanoparticles have been broadly used as antimicrobial agents in food packaging purposes [4]. However, biodegradable films with incorporation of silver nanoparticle products based on one types of biofilms, possess many disadvantages, mainly attributed to brittle nature of films and poor mechanical properties. In order to improve mechanical properties, these films can be modified by several methods such as blending with other natural polymers such as polylactic acid (PLA) and chitosan [5]. Thus, the aim of this research was to optimize mechanical properties of chitosan/PLA/AgNPs. For optimization purpose, RSM approach was applied where a set of experiment was designed using CCD.

2 Materials and Method

2.1 Materials

Chitosan flakes from crab shells were purchased from Fisher Scientific, polyethylene glycol (PEG) 400, polylactic acid (PLA) (pellet form), chloroform and acetic acid were supplied from Sigma Aldrich. Silver nanoparticles (AgNPs) were obtained from our lab. The preparation of silver nanoparticles from tea extract solution was described in previous paper [6].

2.2 Film preparation

Method of chitosan/PLA blend film were prepared as described by [7] with some modifications. To prepare blend films of chitosan and PLA, solution of chitosan and PLA were mixed separately in different solutions before blending. One percent (*w/v*) PLA was dissolved in chloroform and 1% (*w/v*) chitosan was dissolved in 1% (*v/v*) acetic acid and both of them were stirred under magnetic stirring for 2 hours respectively. The various concentration of PEG 400 (in % *w/w*) were then added into both solutions separately and stirred for 1 hour. Then both solutions were blended homogeneously with silver nanoparticles colloids solution. The concentration of silver nanoparticles in the film was fixed at 23.08 % *w/w* based on the finding from the antibacterial test as reported in our previous paper [6]. The various compositions ratios percentage volume of chitosan/PLA and concentration of PEG 400 were listed in the Table 1. The preparation was filtered and degassed under room pressure (101.325 kPa). The blends films were prepared by casting on glass Petri dishes (1-*mm* thickness) and evaporated at 35°C for 18 hours.

2.3 Tensile measurement

The analysis of mechanical properties of the blended films was carried out by using tensile machine (Shimadzu Model Autograph AG-X instrument) according to ASTM D882 standard [8]. Tensile strength (MPa) and elongation at break (%) were involved in this mechanical analysis. Five films were stored at 25°C for 24 hours before measurement were conducted. The analyzed area of films (gage length=50 mm) were unaxially stretched at a fixed velocity of 500 mm/min which the width and thickness of the films were measured before testing. The results of the experiment were computer recorded.

2.4 Experimental design by Central Composite Design (CCD)

Central composite design was used determine the optimum factors and to analyse the response pattern of mechanical properties of biofilms. The range selected for the concentration of polyethylene glycol

(PEG) and percentage volumes of PLA/chitosan were determined according to the restriction that previously experienced during screening the effects and interactions of operational parameters on the films using full factorial design (FFD) [9]. The independent variable for the concentration of polyethylene glycol (PEG) is represented by variable *A* ranging from 5 to 25 %, while the percentage volumes of PLA/chitosan *B* ranged from 20 to 80%, respectively. Based on computer simulation, a total of 12 runs were established including 1 replicate of factorial design, 4 replicate central point and 1 star points ($\alpha = 1.414$).

3 Results and Discussion

3.1 Design of experiment by CCD

Table 1 shows the levels and ranges of analysis process parameters (A: concentration of PEG and B: percentage volume of PLA/chitosan) that influencing the responses (mechanical properties) and the responses results of observed in the experiment.

Table 1. Experimental layout and results of 2^2 full factorial CCD

Factors	**Variables**	**(-α)**	**(-1)**	**(0)**	**(+1)**	**(+α)**
A	Concentration PEG (% w/w)	5.0	7.93	15.0	22.07	25.0
B	Percentage volume of PLA/chitosan	20/80	28.79/71.21	50/50	71.21/28.79	80/20

Standard order	**Run order**	**A**	**B**	**Tensile strength (MPa)**		**Elongation at break (%)**	
				Actual	**Predictd**	**Actual**	**Predicted**
1	8	-1	-1	8.3151	7.82	32.1486	31.40848
2	4	+1	-1	6.0154	6.25	15.1241	13.9999
3	7	-1	-1	3.4066	2.61	8.1466	10.15605
4	1	+1	+1	2.8734	2.80	11.1567	12.78207
5	2	-α	0	4.5648	5.36	25.1411	24.42689
6	3	+α	0	4.6078	4.38	14.1451	13.97406
7	12	0	-α	7.8612	7.93	21.4156	22.91721
8	5	0	+α	1.3093	1.81	9.4152	7.028337
9	10	0	0	5.9431	6.74	13.7415	16.3802
10	9	0	0	7.1245	6.74	15.1781	16.3802
11	11	0	0	8.0126	6.74	20.1561	16.3802
12	6	0	0	5.8671	6.74	16.4451	16.3802

In this design, the quadratic model for predicting the optimal point was according to Eq. 1.

$$Y = b_0 + \Sigma b_i x_i + \Sigma b_{ii} x_i^2 + \Sigma b_{ij} x_i x_j \quad (1)$$

where Y is the response variable, b is the regression coefficient of the model, and x is the coded levels of the independent variables. The Design Expert software version 7.1.6. (Stat-Ease Inc., Minneapolis, USA) was used to develop the experimental plan and optimize the regression equation (Eq.1).

The multiple regression equations describe the mechanical properties of films response using concentration of polyethylene glycol (PEG) and percentage volume of PLA/chitosan as the main variable in terms of coded variables are shown by Eq. 2 and Eq.3:

$$\textbf{Tensile Strength} = +6.736825 - 0.34651A - 2.16454B + 0.441625AB - 0.93355A^2 - 0.93408B^2 \quad (2)$$

$$\textbf{Elongation at Break} = +16.3802 - 3.69564A - 5.61757B + 5.00865AB + 1.410138A^2 - 0.70371B^2 \quad (3)$$

The empirical model equation Eq.1 forms a mathematical correlation model that can be employed to predict and optimize the tensile strength and elongation at break of films within the range of variable factors of this experiment. The response values determined by means of empirical models were compared to the experimental data designed in Table 1 as predicted value. The results are shown in Table 1 for both the tensile strength and elongation at break responses. The response models (RS Model) show good fits to the experimental data. Therefore, the model can be considered adequate for the predictions and for further investigation of optimization purpose.

3.2 Response surface plot

In order to investigate the effects of the parameter on mechanical properties of films, the three dimensional and contour plots were used. Figure 1 depict the interaction between concentration of polyethylene glycol (PEG) (A) and percentage volume of PLA/chitosan (B) for response tensile strength in the form of three-dimensional display of the response surface, contour plot and interaction graph.

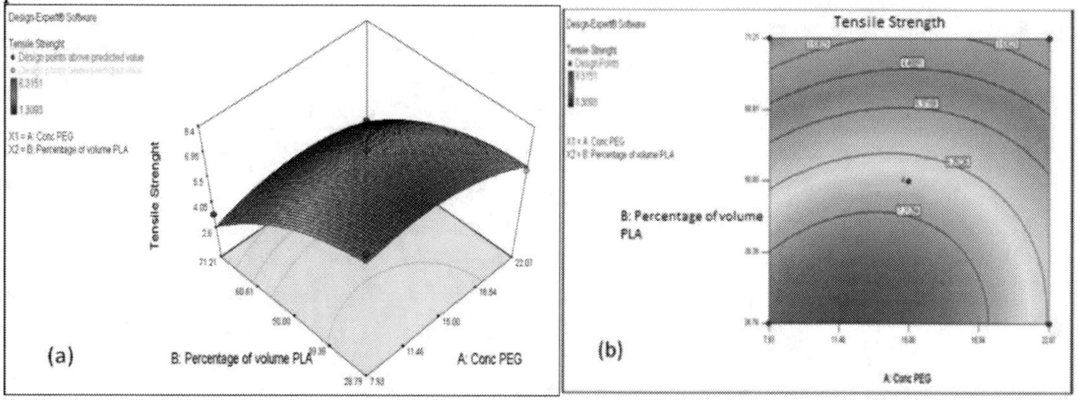

Figure 1. Interactions of polyethylene glycol (PEG) concentration (A) and percentage volume of PLA/chitosan (B) on response tensile strength by (a) response surface curve, (b) contour plot graph

The effect of the polyethylene glycol (PEG) concentration and percentage volume of PLA/chitosan on the tensile strength of films was depicted in Figure 1. This figure demonstrates that the tensile strength of films have reached the maximum value (8.32 MPa) at (7.93 %) concentration of polyethylene glycol (PEG) and (28.79%) percentage volume of PLA/ (71.21%) chitosan. The curvature of the Figure 1 (a) indicated that both concentration of polyethylene glycol (PEG) and percentage volume of PLA/chitosan had quadratic effect on the tensile strength.

Figure 2 illustrates the interactions between concentration of polyethylene glycol (PEG) (A) and percentage volume of PLA/chitosan (B) for response elongation at break in the form of three-dimensional display of the response surface, contour plots and interaction graphs. The effect of the concentration of polyethylene glycol (PEG) and percentage volume of PLA/chitosan on the elongation at break of films was depicted in this figure. This figure validates that the elongation at break of films

 DOI: 10.1051/matecconf/201710901001

have reached the maximum results (32.15 %) at (7.93 %) concentration of polyethylene glycol (PEG) and (28.79 %) percentage volume of PLA/ (71.21 %) chitosan.

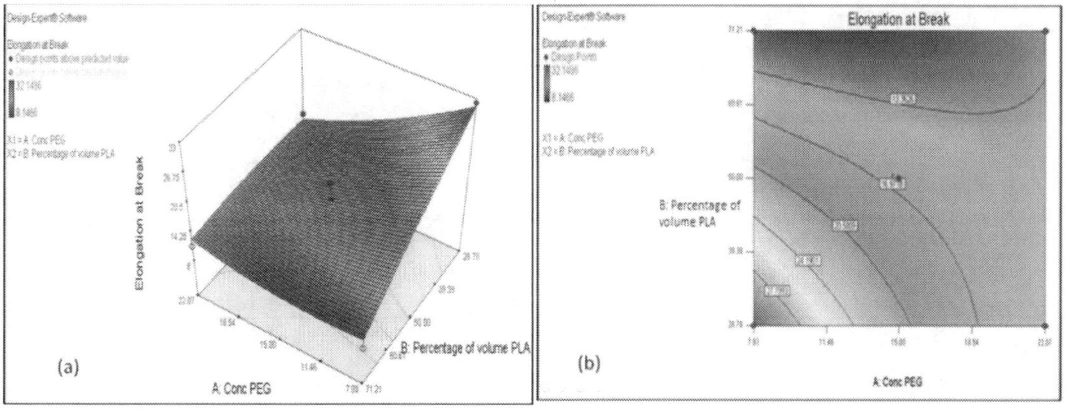

Figure 2. Interaction of concentration of polyethylene glycol (PEG) (A) and percentage volume of PLA/chitosan (B) on response elongation at break by (a) response surface curve, (b) contour plot graph

There was no curvature in Figure 3 (a) indicates that both concentration of polyethylene glycol (PEG) and percentage volume of PLA/chitosan had no quadratic effect on the tensile strength. As discussed in the literature review regarding polyethylene glycol (PEG), as plasticizer, polyethylene glycol (PEG) reduced the intermolecular forces and resulted in reduced tensile strength but increased in percentage value of elongation [10]. Moreover, increasing of polylactic acid (PLA) in the films induced in order to decrease the tensile strength and elongation at break. This is due to the increment of polylactic acid (PLA) concentration, the content of the unreacted monomer and the oligomer also increased, resulting in the deformation of the strong hydrogen bond between the chitosan chains, thus the tensile strength and elongation at break of crude films decreased [11].

3.3 Validation and confirmation run of empirical model adequacy

It must be identified that in a multi-response situation, the excellent condition is when each individual desirability function and the overall desirability D is unity (=1). Note that if an individual desirability function is completely undesirable, then the overall desirability value is zero [12]. From the software results, with overall desirability of D=0.949 (near the perfect value of 1), the obtained optimal conditions are 7.93%w/w and 28.79% for concentration of polyethylene glycol (PEG) and percentage volume of PLA/chitosan, respectively.

To check the prediction accuracy, the empirical model needs to be validated. This could be achieved by the developing of regression equation in predicting the mechanical properties of films at any particular concentration of polyethylene glycol (PEG) (%w/w) and percentage volume of PLA/chitosan (%) within the range of level defined previously. The confirmation runs were carried out by using optimal conditions as suggested by the software (Table 2). Results of Table 2 have shown that the percentage errors are ranging from 2.08% for tensile strength and 3.89% elongation at break, respectively. Thus implied that the empirical model developed were considerably accurate for responding terms as the percentage error between the actual and predicted values were well within the value of 5% [13]. By this means further analysis with regards to ideal operational process for optimal mechanical properties of film would be based on this developed model.

Table 2. The results of verification process

Run Factor as suggested by the software		**Tensile strength (MPa)**			
Concentration of polyethylene glycol (PEG) (%w/w)	Percentage volume of PLA/chitosan (%)	Actual	Predicted	Residual	Error
7.93	28.79	7.99	7.82	0.17	2.08 %
		Elongation at break (%)			
		Actual	Predicted	Residual	Error
		32.6	31.41	1.27	3.89%

4 Conclusions

Data from the optimization process indicates that the optimal conditions for mechanical properties of films were 7.93% w/w concentration of polyethylene glycol (PEG) and 28.79%/71.21% percentage volume of polylactic acid (PLA)/chitosan. At this optimum stage, 7.99 MPa of tensile strength and 32.6 % elongation at break were obtained. Results of verification process have shown that the percentage errors are ranging from 2.08% for tensile strength and 3.89% elongation at break, respectively. Thus implied that the empirical model developed were considerably accurate for responding terms as the percentage error between the actual and predicted values were well within the value of 5%.

Acknowledgment

The authors wish to express their thanks to University Malaysia Pahang and Ministry of Education Malaysia for sponsoring Aznizan postgraduate's study via MyBrain15.

References

1. T.M. Wu, C.Y. Wu, Polym. Degrad. Stab. **91** (2006).
2. M. Aider, LWT - Food Sci. and Tech. **43** (2010).
3. W.Guiga, S. Galland, E. Peyrol, P. Degraeve, A. Carnet-Pantiez, I. Sebti, Inno. Food Sci. and Emerg. Technol. **10** (2009) .
4. R. Yoksan, S. Chirachanchai, Mat. Sci. Eng. C. **30**, 6 (2010).
5. E.Salleh, I.Muhamad, N. Khairuddin, Asian Chitin Journal **3** (2007).
6. A. Shaari, M.N.A. Seman, C.K.M.F.C. Ku, World Research & Innovation Convention on Engineering & Technology 2012, Kuala Lumpur, Malaysia, 3-5 December 2012.
7. F.Sebestian, G. Ste, A. Copinet, V. Coma, Carb. Pol. **5,** 2 (2006).
8. J.W. Vernice, A. Mayer, S.J. Bailey, N.C. Baldini, S. Emery, J. Ermigiotti, B. Huffman, M.A. Olcese, K.A. Peters, J.L. Rosiak, D.A. Terruso, E.A. Whealen, "ASTM Packaging Standards and Related Technical Material 7th Edition," ASTM International, West Conshohocken, PA, 2007.
9. A. Shaari, M.N.A. Seman, C.K M.F.C. Ku, Proceeding of International Conference on Science, Technology and Social Science, Kuantan, Pahang, 20-22 November 2012; 757-765.
10. P.C. Srinivasa, M.N. Ramesh, Tharanathan, Food Hyd., 21, 7 (2007) .
11. F.Yao, W. Chen, H. Wang, H. Liu, K. Yao, P. Sun, H. Lin, Polymer, **44**, 21 (2003) .
12. M. Khayet, M.N. Abu Seman, N., Hilal, J. Mem. Sci. **349** (2010) .
13. A.W. Zularisam, A.F. Ismail, M.R. Salim, M. Sakinah, T. Matsuura, Sep. Pur. Technol. 65, 2 (2009).

Microstructural and Mechanical Properties of Porous 60NiTi Prepared by Conventional Press-and-sinter Method

Khashayar Khanlari[1,a], Maziar Ramezani[1], Muhammad Hayat[2], Kelly Piaras[3], Peng Cao[2] and Thomas Neitzert[1]

[1]Department of Mechanical Engineering, Auckland University of Technology, New Zealand
[2]Department of Chemical and Materials Engineering, The University of Auckland, New Zealand
[3]Department of Engineering Science, The University of Auckland, New Zealand

Abstract. An intermetallic nickel-titanium alloy, 60NiTi, comprised of approximately 60 wt.% Ni and 40 wt.% Ti, contains a broad combination of physical and mechanical properties such as high hardness, low elastic modulus, resistance to aqueous corrosion and good biocompatibility. These unique combinations make this alloy an attractive candidate for medical components such as implants and prosthesis, where biocompatible materials with high hardness and low stiffness are typically used. The conventional press-and-sinter method which represents the least complex, most flexible and economic powder metallurgy method was used to produce porous 60NiTi parts suitable for biomedical applications. The effect of sintering holding time on the microstructure and mechanical properties is investigated. The structure of the as sintered parts is quite porous which is beneficial based on the medical point of view. The ultimate compressive strength of the samples is higher than that of the compact human bone and can, therefore, meet the strength demand of implants for general bone replacement applications.

1 Introduction

An intermetallic nickel-titanium alloy, 60-NiTiNOL, composed of approximately 60 wt.% Ni and 40 wt.% Ti contains a broad combination of physical and mechanical properties such as high hardness, low elastic modulus, resistance to aqueous corrosion and good biocompatibility [1-5]. These unique combinations make this alloy an attractive candidate for medical components such as implants and prosthesis where hard and biocompatible materials with low stiffness are typically employed [4].

Implants for bone replacement in medicine should have a structure similar to the organic tissue to be replaced [4]. The ultimate compressive strength of the bone is less than 340 MPa [6], so the implants, which are going to be used for general bone replacement applications, are expected to show higher or at least 340 MPa strength.

The porous structure is beneficial for bone ingrowth and reduces the Young's modulus of the implant, which helps in mitigation of stress shielding phenomenon [4, 7, 8].

Powder metallurgical (PM) processing is an attractive way to produce porous 60NiTi objects of specific shape with adapted properties. In addition, powder metallurgy methods reduce the machining cost since they can provide near-net-shape components [9].

[a] Corresponding author: kkhanlar@aut.ac.nz

Conventional press-and-sinter (P&S) represents the least complex, most flexible and economic powder metallurgy method available which also results in parts with good dimensional integrity [5, 10]. Due to these attractive reasons, this processing method is used to manufacture the parts in this research. The effect of sintering holding time on the microstructure and mechanical properties is investigated and the obtained results are compared with the prerequisites of bone replacement implants.

2 EXPERIMENTS

In the following section, the applied parameters to produce the samples by conventional press & sinter and the experiments done for obtaining the results are explained.

2.1 Materials

Table 1 lists the relevant information on the starting powders used in this study. The mean particle sizes for Ti and Ni powders are 27.2 μm and 12.57μm respectively and the morphologies of these powders are shown in Fig. 1. Ti powders are spherical while Ni powders have spiky, needle like texture, which is ideal for powder metallurgy applications.

Table 1. Particle size and impurity content of the starting powders

Powder	Particle size (μm)			O(wt.%)	C(wt.%)	N(wt.%)
	D_{10}	D_{50}	D_{90}			
Ti	11.74	27.2	49.04	0.111	0.004	0.047
Ni*	4.33	12.57	42.50	0.100	0.066	0.021

*Ni powders is produced by carbonyl gas process.

Figure 1. Morphologies of the starting powders (a) Ti (b) Ni

After characterizing the powders, a batch of powder mixture, i.e. Ni/Ti having a nominal composition of 60 wt.% Ni and 40 wt.% Ti, were gently mixed in a ball mill for 45min with a ball-to-powder weight ratio of 3:1.

2.2 Pressing and sintering

After mixing, powder mixtures were compacted into cylindrical discs (20 mm in diameter and 5 mm-thick) and compression testing cylinders (12 mm in diameter and 20 mm-thick) in a single-action steel die with 400 MPa pressure. Subsequently, the green compacts were sintered in a vacuum furnace at $3*10^{-3}$ Pa.

In the first stage, samples are heated till 700 °C by applying 10 °C/min heating rate. While in the second stage slow heating rate of 2 °C/min, from 700 °C till 900 °C was applied. Finally, in the last stage, 10 °C/min heating rate is used again to reach the 1050 °C. Samples were held at 1050 °C for different holding times of 2 hours, 4 hours and 6 hours and finally furnace cooled with a slow cooling rate. Long term sintering was considered highly impractical in view of any industrial or practical applications for 60NiTi (wt.%), and therefore, in the current study, holding time was varied between 2 to 6 hours only.

2.3 Characterization methods and mechanical tests

Microstructures of the as-Sintered compacts were observed using a scanning electron microscope (Hitachi-su70) equipped with an Energy dispersive X-ray spectrometer (EDX). Scanning Electron Microscope (SEM) images were taken from the cross section of the samples in different magnifications. Standard metallographic procedures were used to prepare the specimen for microscopy, starting with rough grinding using a P220 (European scale) SiC paper and continuing with P360, P800 and P1200 at 300 rpm concluding with vibratory polishing with 3 μm colloidal silica. Finally, the grinded and polished samples were swab-etched with a room temperature aqueous solution of 1 vol.% HF and 10 vol.% HNO3 for 86 seconds [11]. X-ray Diffraction (XRD, Bruker D2 Phaser) was used to identify the generated phases after sintering.

2.4 Mechanical tests

The compressive properties of the 20-mm-thick samples compressed after 1050 °C sintering for different holding times were measured on an MTS 810 universal machine with a displacement rate of 0.01mm/s at room temperature. An alignment cage ensured the parallelism of all samples during testing. The ends of compression cylindrical samples (machined into 5-mm diameter and 7.5-mm height) were polished and smoothed using sand papers, and finally the ends were greased before compression tests.

3 Results and discussion

Results of the microstructural analysis, phase identification and mechanical tests are discussed in the following section.

3.1 Microstructural analysis and phase identification

Phase constituents were identified by using X-ray Diffraction (XRD, Bruker D2 Phaser). Based on the XRD results (Fig.2) austenitic B2 NiTi phase dominates with the presence of minor martensitic B19, intermetallic $NiTi_2$, Ni_3Ti and Ni_3Ti_2 (metastable phase) phases [12]. There is no evidence of Ni or Ti elements in XRD patterns. Pure Ni is carcinogenic and its release in body environment can be harmful [13]. Thus, from the medical standpoint, sintering process has been successful to vanish pure Ni.

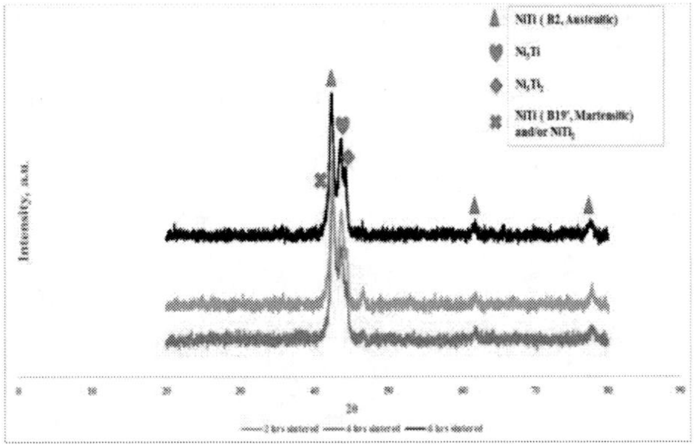

Figure 2. XRD patterns for the sintered 60-NiTi.

SEM images of samples compacted at 400 MPa and sintered at 1050°C for different holding times of 2 hours, 4 hours and 6 hours can be observed in Fig.3. As it can be seen from the SEM micrographs, samples are quite porous, and the majority of pores are interconnected. This is beneficial for bone ingrowth and in addition results in the decrease of stiffness. Decrease in stiffness helps prevention of stress-shielding phenomenon, which is so common in implants and often leads to implantation failure.

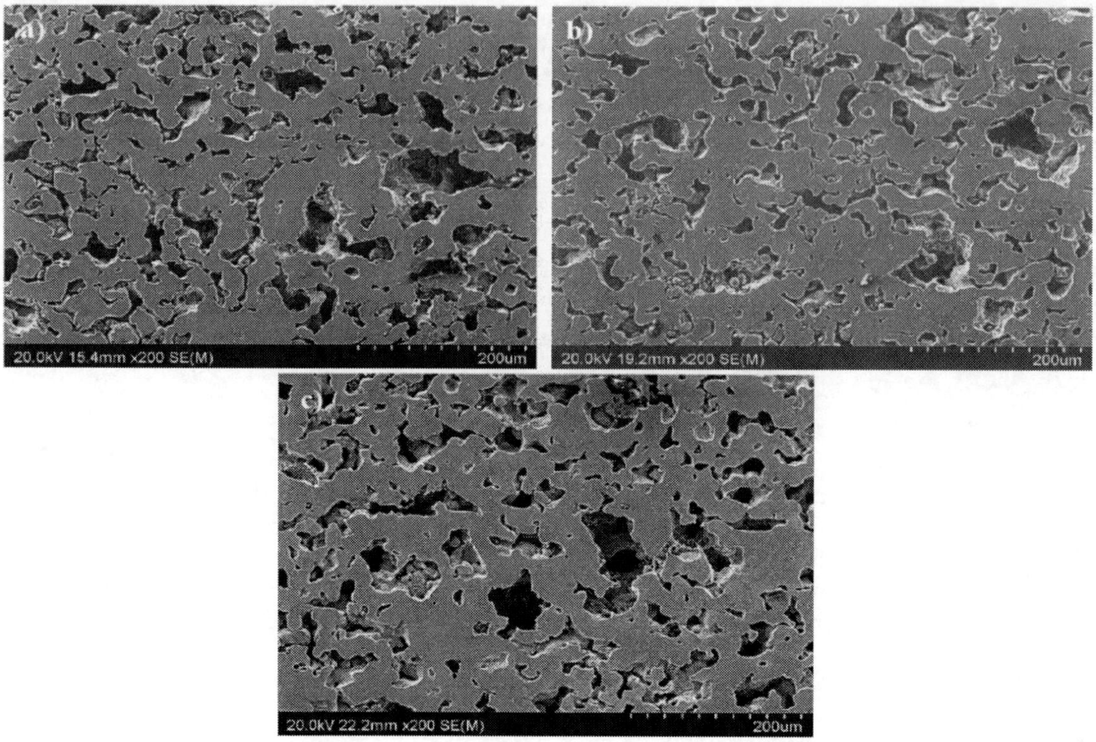

Figure 3. SEM images of 400 MPa compacted samples after sintering at 1050°C C and various holding times (a) 2 hours holding time (b) 4 hours holding time (c) 6 hours holding time

3.2 Mechanical tests

Fracture strength of the sintered samples were measured by applying compression tests. Results of the stress-strain fracture tests can be seen in Fig.4. As it can be observed, all the samples fracture in brittle manner. This can be attributed to the porous structure of the material and existence of sharp pores, which result in stress concentration and local deformation [8]. Furthermore, the increased instantaneous stiffness after linear deformation stage is larger than that of dense materials. Li et al. attributes this phenomenon to the decrease of porosity during compression test and the fact that there is stress concentration on the edge of pores [14].

In addition, the ultimate compressive strength (Fig.5) of the samples is higher than 340 MPa, which is higher than that of the compact human bone and as a result can meet the strength demand of bone implants [6]. The weakest portion in the material rules the strength of a material. In this regard, porosity, pore size and pore shape have significant influence on the strength of porous 60NiTi alloys [7, 8, 15].

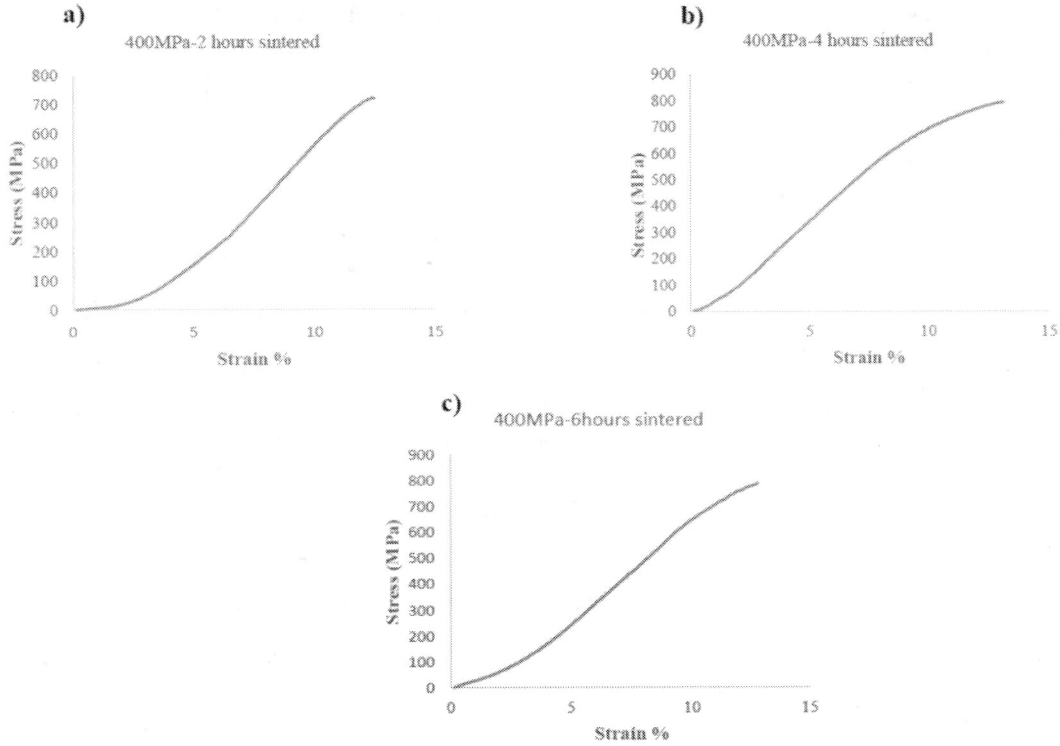

Figure 4. Fracture strength of 400 MPa compacted samples after sintering at 1050°C and various holding times (a) 2 hours holding time (b) 4 hours holding time (c) 6 hours holding time

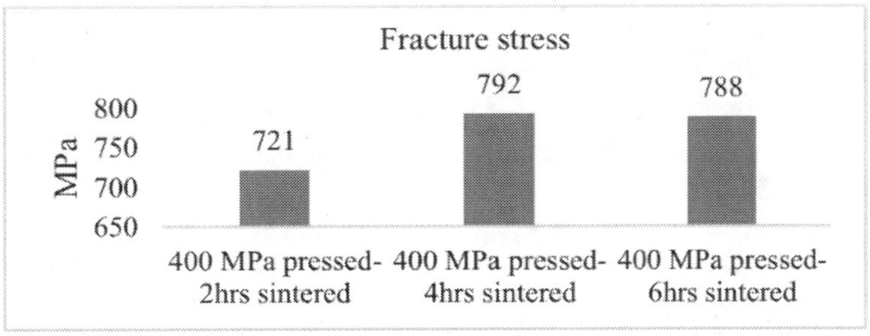

Figure 5. Ultimate fracture stress of the samples under compassion test.

In this stage, the porosity, due to the shrinkage of the pores, decreased with the increase of holding time. As a result, fracture strengths improvement is due to internal porosity reduction, which can result in decrease of the local stress concentrations [15]. However, fracture strength of 6 hours sintered sample is slightly less than the 4 hours sintered one. This means that the mechanical properties of the prepared porous 60NiTi alloy were mainly influenced by the size of the pores and their sharpness in the sample, but not the porosity. As it can be seen in Fig.3, 6 hours sintered samples have sharper pores compared to 4 hours sintered ones. Stress concentration emerges easily at the sharp edges of the pores, which results in the decrease in the strength of the porous materials [8].

4 Conclusions

In this study we synthesized porous 60NiTi alloys from Ni/Ti powder mixtures using a conventional press-and-sinter method. The effects of sintering holding time (2 hours, 4 hours and 6 hours) on mechanical strength of the sintered samples were investigated. Satisfactory results were obtained in all the cases. Results show that pure carcinogenic Ni has been successfully vanished during sintering process. In addition, the structure of produced parts is quite porous which is good for bone ingrowth. The ultimate compressive strength of the samples is higher than that of the compact human bone, which is lower than 340 MPa. Therefore, the produced samples meet the strength demand of bone implants.

References

1. Dellacorte, C., et al., Intermetallic nickel-titanium alloys for oil-lubricated bearing applications. 2009.
2. DellaCorte, C., et al., Resilient and corrosion-proof rolling element bearings made from superelastic Ni-Ti alloys for aerospace mechanism applications, in Rolling Element Bearings. 2012, ASTM International.
3. Buehler, W.J. and F.E. Wang, A summary of recent research on the Nitinol alloys and their potential application in ocean engineering. Ocean Engineering, 1968. 1(1): p. 105IN7109-108IN10120.
4. Bansiddhi, A., et al., Porous NiTi for bone implants: a review. Acta Biomater, 2008. 4(4): p. 773-82.
5. Sadrnezhaad, S. and S. Hosseini, Fabrication of porous NiTi-shape memory alloy objects by partially hydrided titanium powder for biomedical applications. Materials & Design, 2009. 30(10): p. 4483-4487.
6. Li, H., et al., High-porosity NiTi superelastic alloys fabricated by low-pressure sintering using titanium hydride as pore-forming agent. Journal of materials science, 2009. 44(3): p. 875-881.
7. Chen, G., P. Cao, and N. Edmonds, Porous NiTi alloys produced by press-and-sinter from Ni/Ti and Ni/TiH 2 mixtures. Materials Science and Engineering: A, 2013. 582: p. 117-125.
8. Zhu, S., et al., Stress–strain behavior of porous NiTi alloys prepared by powders sintering. Materials Science and Engineering: A, 2005. 408(1): p. 264-268.
9. Bertheville, B., M. Neudenberger, and J.-E. Bidaux, Powder sintering and shape-memory behaviour of NiTi compacts synthesized from Ni and TiH 2. Materials Science and Engineering: A, 2004. 384(1): p. 143-150.
10. Cluff, D. and S. Corbin, The influence of Ni powder size, compact composition and sintering profile on the shape memory transformation and tensile behaviour of NiTi. Intermetallics, 2010. 18(8): p. 1480-1490.
11. Stanford, M.K., Hot Isostatic Pressing of 60-Nitinol. 2015.
12. Hornbuckle, B.C., et al., Hardening behavior and phase decomposition in very Ni-rich Nitinol alloys. Materials Science and Engineering: A, 2015. 639: p. 336-344.
13. Li, B.-Y., L.-J. Rong, and Y.-Y. Li, Porous NiTi alloy prepared from elemental powder sintering. Journal of materials research, 1998. 13(10): p. 2847-2851.
14. Li, B.-Y., L.-J. Rong, and Y.-Y. Li, The influence of addition of TiH 2 in elemental powder sintering porous Ni–Ti alloys. Materials Science and Engineering: A, 2000. 281(1): p. 169-175.
15. Sadrnezhaad, S.K. and O. Lashkari, Property change during fixtured sintering of NiTi memory alloy. Materials and manufacturing processes, 2006. 21(1): p. 87-96.

Impact Strength of Natural Fibre Composites Measured by Different Test Methods: A Review

Namasivayam Navaranjan[1,a] and Thomas Neitzert[2]

[1]*School of Applied Sciences and Mathematics, Universiti Teknologi Brunei, Jalan Tungku Link, Gadong BE1410, Brunei Darussalam*
[2]*School of Engineering, Computer and Mathematical Sciences, Auckland University of Technology, Private Bag 92006, Auckland 1142, New Zealand*

Abstract. Different types of impact test methods have been used in recent years to measure the impact resistance of natural fibre composites (NFCs). After reviewing the literature, the impact resistance of flax, hemp, sisal, wood and jute fibre composites that were measured using different test methods have been compared and discussed. It has been learned that the test methods were selected for research interest, industry requirement or availability of test equipment. Each method had its own advantages and limitations. The result from a particular test could be compared but not with the result from other test methods. Most impact test methods were developed for testing ductile-brittle transition of metals. However, each NFC has a different morphology and cannot be comparable to metals in failure mode and energy absorption characteristic during an impact test. A post evaluation of morphology of an NFC sample after an impact test is important to characterise the material.

1 Introduction

Natural fibre composites (NFC) are made of a polymer matrix reinforced with cellulose-based plant fibres such as flax, hemp, sisal, wood and jute. They are receiving increasing attention in automobile, building and other engineering applications. The composites are cost effective, and have low density, high specific strength and disposability. Furthermore, most of NFCs are recyclable while the natural fibre and biopolymer composites can be biodegradable [1]. NFCs have good electrical and chemical resistance. They also possess good thermal and acoustic insulating properties.

Despite many advantages of NFCs, their impact strength is low compared to synthetic fibre composites [2]. Impact resistance is the ability of a material to withstand a shock loading or an applied stress at high speed. Impact behaviour is an important mechanical property of engineering materials that are used for many popular applications including interior and exterior components of automobiles, buildings, aircrafts and many more. It depends on a number of parameters such as strength, elastic modulus, length and orientation of fibres, and fibre-matrix interfacial bond strength. It also depends on the type of test method used to measure the impact energy.

There are two categories of impact tests: low velocity impact, and high velocity impact [3]. Charpy, Izod and drop weight impact tests fall into the first category while the ballistics impact test falls into the second category. Charpy and Izod impact test methods are also known as pendulum method and

[a] Corresponding author: nava.navaranjan@utb.edu.bn

are simple and easy to perform [4]. The method determines the amount of energy absorbed by notched or unnotched sample of a material during fracture. The absorbed energy is a measure of a given material's toughness and acts as a tool to study the ductile-brittle transition temperature.

Drop weight impact tests are generally used to test the impact behaviour of composite plates and components [5]. This test can closely resemble impact damage that could occur during in-service. Two categories of damages can be observed after a drop weight impact test. The first category is macro-damages that can be visible to the naked eye and the second is at microscale that can seldom be seen by the naked eye. Evaluation of both types of damage can be enhanced through the use of post-impact testing.

Low velocity impact test instruments can be modified to be sophisticated and instrumented impact test devices. These devices have three major components: the dynamic load cell (called tup), the data display system, and the signal processing unit [6]. The tup is placed on the impactor used to strike the specimen. The tup consists a strain gauge that measures the change in strain vs. time as the impactor strikes the specimen. The measured load and strain against the time can provide impact behaviour of a sample from the initiation to the end of failure.

The ballistics test is used to evaluate the impact resistance of composite panels and products such as composite armor [7]. The test involves the firing a high speed projectile at a product of interest and determining the localized the damage after the impact.

A review of impact test methods that are commonly used for NFCs are briefly presented in this paper. The impact resistance of some flax, hemp, sisal, wood and jute fibre composites that were measured using different test methods are compared and discussed. Finally, the benefits and limitations of using each method for NFCs are presented.

2 Impact test method

2.1 Pendulum method

In this method, a pendulum of a known weight is hoisted to a known height and allowed to fall and strike a specimen clamped at the bottom of the swing. The pendulum continues to swing up after breaking the specimen to a height somewhat lower than that of a free swing. The energy lost by the pendulum is measured as the impact energy of the specimen. Izod and Charpy use the same principle to measure the impact resistance [5]. The difference in both methods is that in the Charpy impact test the specimen is supported as a simple beam, and in the Izod impact test the specimen is supported as a cantilever. They also have different test procedure and specimen geometry as given in Table 1.

Table 1. Izod and Charpy test requirements.

Procedure	Izod	Charpy
Common standards	ASTM D256 or ISO 180 for notched specimens ASTM D4812 ISO 180 unnotched specimens	ASTM A370, ISO 148 or EN 10045-1
Test sample	Placed in vertical position	Placed in horizontal position
Test hammer striking	At the upper tip of specimen	At point of notch but in opposite direction
Notch face	Facing the striker, fastened in a pendulum	Face is positioned away from the striker
Notch type	V-notch	V and U notches
Specimen dimensions	75 x 10 x 10mm	55 x 10 x 10mm
Hammer type	Farming hammer as striker	Ball Pin hammer as striker

2.2 Drop weight impact test

The test is performed according to ASTM 7136. A known weight is dropped in a vertical direction and the impact energy is calculated from the dropped height. Since the falling weight either stopped on the test specimen, or destroyed the specimen completely and passed through it, the only result that could be obtained is of a pass or fail the test [8]. Falling weight impact have the following advantages over other methods:

- It is used for samples of moulded parts, plates and sheets.
- Failures originate at the weakest point in a sample and propagate from there. The failure propagation has no preferential direction.
- Samples don't have to shatter to be considered failures. Failure can be defined by deformation, crack initiation, or complete fracture, depending on the requirements.

2.3 Instrumented impact test

Instrumented impact test devices are made with electronic sensing instrumentations that can record the load on the test specimen continuously as a function of time and specimen deformation prior to fracture. The instruments can record strain or stress against time for the entire period of an impact test. The data from the acquisition system provides complete representation of impact test history rather than a value from a single calculated number. Another advantage is that the test times can be reduced because of its automated data acquisition system [9]. Instrumented drop weight and pendulum test equipment are considered to be the best general impact tests devices presently available. A very complete impact profile can be developed for an NFC using multiple tests at various rates. This approach is useful for simulation and comparison of NFC's functional impact resistance. There is enough flexibility to simulate real life conditions from multiple test data and perform audit inspections on samples, parts or moulded components.

2.4 Ballistics test

The European EN 1522 / 1523 and ASTM E3062 standards describe the Ballistics test. It is a high speed testing that is useful for the ultimate impact strength of composites. The ballistics testing is done by firing an impactor traveling in the range of 400-2000 m/s [10]. The Ballistics method consists of firing a high speed projectile at an object and determining after the impact how localized the damage is. Structural response of a material is less important in high velocity impact than in a low velocity condition. The damage area is more localized by the impact and the geometrical consideration is less important. This is a good method for testing impact resistance of natural fibre composites, and has been used for testing products such as composite panels, armors and windscreens of aircraft.

3 Impact strength of natural fibre composites

The impact strength of some natural fibre composites of flax, hemp, sisal, wood and jute are given in Table 2. The type of matrix and the impact test method used for each composite are also listed in Table 2.

Table 2. Impact strength of selected natural fibre composites

Matrix/process	Fibre/type of orientation	Fibre Content (%)	Izod Test (kJ/m^2 or J/m)	Charpy Test (kJ/m^2 or J/m)	Drop Test	Ballistics Test	Reference
PP	Flax yarn (aligned) Pultruded flax	30		88			[11]
PP/CM	Flax (aligned) needle punched flax/PP mats	50	751				[12]
UP/ RTM	Flax (random)	40		33			[13]
PP/ Rubbery MAPP coupled IM	Flax	30		22			[14]
PP	Flax IM	30		18			[15]
Vinyl ester	Flax mat	30			Force = 3.5 kN Energy = 65%		[16]
Epoxy/CM	Hemp (random)	65		15			[17]
PLA/CM	Hemp (carded) alkali treated	30		9			[18]
PLA/CM	Hemp (aligned) wrap spun alkali treated short hemp hybrid yarn	30		19			[19]
PLA/CM	Hemp (biaxial) wrap spun bleached hemp hybrid yarn.	45		25			[20]
PLA/CM	Hemp (DSF)	30		11			[17]
PP/MAPP coupled IM	Hemp	40	210				[21]
PLA	Sisal (random)	30	3.3				[22]
PLA/PEG-plasticized	Sisal	30	4.5				[22]
Epoxy	Sisal (unidirectional)	30	22				[23]
UFR	Sisal (random)	50		9.5 (un-notched)			[24]
Epoxy	Sisal (aligned)	30				Energy = 106 J	[25]
PER/MEK	Curaua (aligned)	30		140			[26]
PP	Wood (flour)	30	5				[24]
PP/EPR/MAPP	Wood (flour)	30	7				[24]
PP/MAEPDM	Wood (flour)	30	7				[24]
PP/MAPP	Wood (flour)	30	5				[24]
PP/ MAPP coupled IM	Wood (BKP)	40	40				[27]
PP	Wood, TMP (fibre length = 3.0 mm)	40	5.5				[28]
PP/MAPP	Wood, MDF	30	4				[29]
Unsaturated polyester	Jute mat	30	6		Initiation energy 8 J, Max energy 13J		[30]
Polyester, 10% NaOH	Jute	18		3.25			[31]
Epoxy	Jute, chopped	30	115				[32]

RTM = Resin transfer moulding; CM = Compression moulding; IM = Injection moulding

PP = poly propylene; EPR = ethylene propylene-diene; PEG = polyethylene glycol; MAPP = maleated PP;
MAEPDM = maleated ethylene-propylene-diene elastomer; PER = polyester resin; MEK = metyl-ethyl-ketone;
UFR = urea-formaldehyde resin
BKP = bleached kraft pulp; PALF = pineapple leaf fibre.
PHB = poly (3-hydroxybutyrate); PLLA = L-polylactide acid; PA = Polyamide.
MAA-PP = maleic acid anhydride modified PP.
CSM = chopped strand mat; DSF = Dynamic sheet forming; TMP = thermo-mechanical pulp; MDF = medium density fibre

4 Comparison of different impact test methods

NFCs have growing interest in engineering, building and furniture applications that require impact resistance. Hence, a comparison of available impact test methods can help in choosing single or combination of correct test methods to obtain reliable and accurate impact properties. Izod and Charpy impact tests are easy and fast methods that allow to generate large amounts of data. However, the results obtained are not in-depth as well as do not reveal a great details about the impact property of material. These methods are cost effective to use as they can easily be set up to test a great number of specimens and generate a large amount of data within a short period of time.

Izod and Charpy methods were initially developed for metals and the test results were commonly used to compare the impact response of the materials. Metals are isotropic materials that are different from NFCs in composition, microstructure and manufacturing condition [33]. These method can miss important information of test specimen behaviour and characteristic of the material during the impact event and can provide misleading information of impact strength. For example, composites can fail internally but display no damage externally. Sophisticated instruments that consist of high speed photography are able to produce results showing the propagation of the crack induced by the impact load. These instruments allow the user to measure the force and/or strain more accurately. The data, however, may not be suitable for some composite materials due to the anisotropic nature of the material [34]. For these composites, results from an impact test have limitation to provide the overall behaviour of the materials. Furthermore, the simplistic nature of impact testing and test results from an NFC have limited post impact test methods that can be used for detail analysis of impact behaviour.

The drop weight impact test method is commonly used to evaluate the impact behaviour of a product or component that is normally from a "real world" application and cannot be tested using the Charpy or Izod instrument. The drop weight method can provide greater amount of information on the behaviour of a composite, for example, characteristics of interfaces of laminated composites [35]. The drop weight impact method also allows different configurations to be used depending on geometry and size of samples. Therefore, drop weight testing tends to be the preferred method when using low velocity impact testing. Drop weight testing results may be enhanced through post impact testing.

Natural fibre laminated composites can be subjected to delamination due to their low intraluminal strength. Shear and normal stresses can occur in the boundary layer of laminates under transverse and normal loading. The polymeric matrix of a composite shares the energy among the constituents and makes the composite to absorb impact energy. A low velocity impact state does not produce perforation, but creates delaminations between the layers with no visible damage on surfaces [36]. A composite may also have pre-existing delamination due to contaminated reinforcing fibres, insufficient wetting of fibres, machining and mechanical loading and the lack of reinforcement in the thickness direction [37]. Such defects can reduce the energy absorbing ability of a composite.

The mechanism and mode of failure of NFC can be affected by a few parameters due to low velocity impact loading conditions, i.e. fibre-matrix interface, type of fibre layup, specimen thickness, velocity and type of impactor [38]. Metals generally undergo elasto-plastic deformation prior to failure. Hence, they absorb energy during both elastic and plastic deformations during an impact test. Composite laminates have negligible plastic deformation and mostly absorb energy during elastic deformation. Since most composites are brittle in nature, they can absorb energy during elastic deformation and each composite can have similar damage mechanisms.

Ballistics is a high velocity test and cannot be directly compared to the Charpy, Izod and drop weight tests. It is complicated since a high velocity impactor is involved in the test. However, the Ballistics test is important when a sample of structural components should be assessed under an impact loading condition that is closer to reality. The Ballistics test is highly effective in replicating the behaviour of the composite components at high speed impact.

References

1. Md. S. Islam, A. Kovalcik, M. Hasan, V. K. Thakur. Natural Fiber Reinforced Polymer Composites. International Journal of Polymer Science. (2015).
2. K.L. Pickering, M.G. Aruan Efendy, T.M. Le. A review of recent developments in natural fibre composites. Composites: Part A **83** (2016) 98–112.
3. R. Zah, R. Hischier, A.L. Leao, I. Braun. Curaua fibers in the automobile industry – A sustainability assessment. Journal of Cleaner Production. **15**(2007) 1032-1040.
4. M. Nagai H. Miyairi. The Study on Charpy impact testing method of CFRP; Advanced Composite Materials: The Official Journal of the Japan Society of Composite Materials. **3** (1994) 177-190.
5. S.N.A. Safri, M.T.H. Sultan, N. Yidris, F. Mustapha. Low Velocity and High Velocity Impact Test on Composite Materials – A review. The International Journal of Engineering and Science. **3** (2014) 50-60.
6. C. Clemons, A. R. Sanadi. Instrumented Impact Testing of Kenaf Fiber Reinforced Polypropylene Composites: Effects of Temperature and Composition. Journal of Reinforced Plastics and Composites. (2007).
7. F. S. da Luza, E. P. L. Juniora, L. H. L. Louroa, S. N. Monteiro. Ballistic Test of Multilayered Armor with Intermediate Epoxy Composite Reinforced with Jute Fabric. Materials Research. **18**(2015).
8. C.V. Srinivasa, K.N. Bharath. Impact and Hardness Properties of Areca Fiber-Epoxy Reinforced Composites. J. Mater. Environ. Sci. **2** (4) (2011) 351-356.
9. A. Malhotra & F. J. Guild. Impact Damage to Composite Laminates: Effect of Impact Location. Appl Compos Mater **21** (2014) 165–177.
10. S. D. Rossoa, L. Iannuccia, P. T. Curtis. On the ballistic impact response of microbraid reinforced polymer composites. Composite Structures **137** (2016) 70–84.
11. I. Angelov, S. Wiedmer, M. Evstatiev, K. Friedrich, M. G. Pultrusion of a flax/ polypropylene yarn. Composites Part A, **38**(5) (2007)1431–8.
12. K. Oksman. Mechanical properties of natural fibre mat reinforced thermoplastic. Appl Compos Mater, **7**(5–6) (2000) 403–14.
13. Z. Li, X. Zhou, and C. Pei. Effect of Sisal Fiber Surface Treatment on Properties of Sisal Fiber Reinforced Polylactide Composites. International Journal of Polymer Science. **2011**(2011).
14. M. H. B. Snijder, H. L. Bos. Reinforcement of polypropylene by annual plant fibers: optimisation of the coupling agent efficiency. Compos Interfaces **7** (2) (2000) 69–79.
15. A. K. Bledzki, A. A. Mamun, M. Lucka, V. S. Gutowsk. The effects of acetylation on properties of flax fibre and its polypropylene composites. Express Polym Lett **2**(6) (2006) 413–22.
16. A. Pavlovic, C. Fragassa, C. Santulli. Experimental Analysis of Mechanical Properties of Composites Reinforced by Flax and Basalt Natural Fibers. 5th Conference on Natural Fibre Composites, Rome 15-16 October 2015.
17. M. S.Islam, K. L.Pickering, N. J. Foreman. Influence of alkali fiber treatment and fiber processing on the mechanical properties of hemp/epoxy composites. J Appl Polym Sci **119**(6) (2011) 3696–707.
18. I. S. Islam, K. L. Pickering, N. J. Foreman. Influence of alkali treatment on the interfacial and physico-mechanical properties of industrial hemp fibre reinforced polylactic acid composites. Composites Part A, **41**(5) (2010) 596–603.

19. B. Baghaei, M. Skrifvars, M. Salehi, T. Bashir, M. Rissanen, P. Nousiainen. Novel aligned hemp fibre reinforcement for structural biocomposites: porosity, water absorption, mechanical performances and viscoelastic behaviour. Composites Part A, **61** (2014) 1–12.
20. I. Van de Weyenberg, J. Ivens, A. De Coster, B. Kino, E. Baetens, I. Verpoest. Influence of processing and chemical treatment of flax fibres on their composites. Compos Sci Technol **63**(9) (2003) 1241–6.
21. M. Sain, P. Suhara, S. Law, A. Bouilloux. Interface modification and mechanical properties of natural fiber-polyolefin composite products. J Reinf Plast Compos **24**(2) (2005) 121–30.
22. M. K. Guptaa, R. K. Srivastava. Properties of sisal fibre reinforced epoxy composite. Indian Journal of Fibre & Textile Research. **41**(2016), 235-241.
23. J. B. Zhong, J. Lv, C. Wei. Mechanical properties of sisal fibre reinforced ureaformaldehyde resin composites. eXPRESS Polymer Letters. **1**(10) (2007) 681–687.
24. G. Keledi, A. Sudár, Ch. Burgstaller , K. Renner, J. Móczó, B. Pukánszky. Tensile and impact properties of three-component PP/wood/elastomer composites. eXPRESS Polymer Letters **6**(3) (2012) 224–236.
25. B. M. de Araújo, E. P. Lima, L. H. L. Louro, S. N. Monteiro. Ballistic Performance of Sisal Fibre Reinforced Epoxy Composite in Multilayered Armor. 21° CBECIMAT - Congresso Brasileiro de Engenharia e Ciência dos Materiais 09 a 13 de Novembro de 2014, Cuiabá, MT, Brasil.
26. A. da S. FerreiraI, F. P. D. LopesI, S. N. MonteiroI, K. G. Satyanarayana. Charpy impact resistance of alkali treated curaua reinforced polyester composites. Revista Matéria, **15**(2) (2010), 131–137.
27. H. J. Li, M. M. Sain. High stiffness natural fiber-reinforced hybrid polypropylene composites. Polym Plast Technol Eng., **42**(5) (2003):853–62.
28. A. Thumm, A. R. Dickson. The influence of fibre length and damage on the mechanical performance of polypropylene/wood pulp composites. Composites: Part A, **46** (2013) 45–52.
29. A. R. Dickson, D. Even, J. M. Warnes, A. Fernyhough. The effect of reprocessing on the mechanical properties of polypropylene reinforced with wood pulp, flax or glass fibre. Composites: Part A, **61** (2014) 258–267.
30. E. Elbadry, H. Hamada. Impact Properties of Natural Jute Fabric/Jute Mat Fiber Reinforced Polymer Matrix Hybrid Composites. Journal of Mechanics Engineering and Automatio, **2** (2012) 381-388.
31. A. Gopinatha, M. S. Kumar, A. Elayaperumal. Experimental Investigations on Mechanical Properties of Jute Fiber Reinforced Composites with Polyester and Epoxy Resin Matrices. 12th GLOBAL CONGRESS ON MANUFACTURING AND MANAGEMENT, GCMM 2014.
32. R. Naveenkumar, V. Sharun, B. Dhanasakkaravarthi, P. T. I. Rajakumar. Comparative study on jute and Kenaf fiber composite material. Int. Journal of Applied Sciences and Engineering Research, **4**(2), 2015.
33. Hufenbach, W., et al. Charpy impact tests on composite structures – An experimental and numerical investigation. Composites Science and Technology, **68** (2008), 2391-2400.
34. Ujihashi, Sadayuki. An intelligent method to determine the mechanical properties of composites under impact loading. Composite Structures, (1993) 149-163.
35. Meola, Carosena, Carlomagno, M. Giovanni. Impact damage in GFRP: New insights with infrared thermography. Composites: Part A (2010), 1839-1847.
36. H. Ghasemnejad, V. R. Soroush, P. J. Mason. To improve impact damage response of single and multi-delaminated FRP composites using natural Flax yarn. Materials and Design, (2012) 865-873.
37. H. Ghasemnejad, A. S. M. Furquan, P. J. Mason. Charpy impact damage behaviour of single and multi-delaminated hybrid composite beam structures. Materials and Design, (2010) 3653-3660
38. C. Scarponi, C. S. Pizzinelli, S. Sanchez-Saez, and E. Barbero. Impact load behaviour of Resin Transfer Moulding (RTM) hemp fibre composite laminates. Journal of Biobased Materials and Bioenergy, **3**(3) (2009), 298-310.

Investigation of the Microstructures, Mechanical Properties and Corrosion Behaviours of Ti-8Mo-6Ni Alloys Adding NbC Powders *via* Vacuum Sintering Process

Cheng Liang[1, 2,a], Shih-Hsien Chang[2], Wei-De Wang[2], Kuo-Tsung Huang[3] and Shun-Tian Lin[1]

[1]*Department of Mechanical Engineering, National Taiwan University of Science and Technology, Taipei 10607, Taiwan, ROC*
[2]*Department of Materials and Mineral Resources Engineering, National Taipei University of Technology, Taipei 10608, Taiwan, ROC*
[3]*Department of Auto-Mechanics, National Kangshan Agricultural Industrial Senior High School, Kaohsiung 82049, Taiwan, ROC*

Abstract. In this study, the Ti-8Mo-6Ni alloys are subjected to add various amounts of NbC carbides as a strengthening phase after the sintering process. Meanwhile, the composite materials underwent a vacuum sintering process at different temperatures of 1175°C, 1225°C and 1275°C for 1 h, respectively. To evaluate the microstructure, mechanical properties and corrosion behaviors of the Ti-8Mo-6Ni alloys adding NbC *via* the vacuum sintering processes, tests on the relative density, hardness, transverse rupture strength (TRS), corrosion and microstructure inspections are performed using XRD and SEM, with the aim of improving the microstructure and properties of sintered Ti-8Mo-6Ni alloys. The experimental results showed that when 1 wt% of NbC carbides is added to the Ti-8Mo-6Ni alloys they possessed the optimal properties after sintering at 1275°C for 1 h. The sintering density increased to about 99.7%, the hardness is enhanced to 48.1 HRC and the TRS increased to 976 MPa. Meanwhile, the sintered composite materials possessed good corrosion resistance.

1 Introduction

During the past few decades, β titanium alloys have become one of the most promising candidates due to their high specific strength and good corrosion resistance [1, 2]. Mo is desirable as an alloying element for the development of Ti-based biomaterials. Previous researches have studied the phase transformations and mechanical properties of different Ti-Mo alloys, and found that the phase constitutions, elastic modulus and hardness are different for Ti-Mo alloys with different Mo contents [3, 4]. It is remarkable that the Ti-8Mo exhibits the highest mechanical properties among the porous Ti-(4-10 wt%)Mo alloys [2, 5]. Powder metallurgy (P/M) methods offer various types of materials with the objective of achieving higher strength, hardness and wear resistance for composite materials such as titanium-based alloys powders. Conventional P/M involves mixing the metal powders, compacting of the mixed powders into moulds and then sintering of the compact powders under the different atmospheres [6, 7]. Furthermore, our previous study has shown that the combination of these

[a] Corresponding author: lcheng@ntut.edu.tw

elements in a Ti-8Mo-6Ni alloy has a good balance between mechanical properties and hot workability [2].

The niobium carbide (NbC) possesses a high melting point (3600°C) and hardness (HV 2800). It is known to inhibit grain growth in high temperature of PM process [8, 9]. Moreover, the wettability with binder of NbC is much better than the other carbides. In recent years, attention of many researchers was taken on NbC for its attractive properties. On the other hand, the previous literature also indicated that the addition of NbC substantially refined the structure and improved the mechanical properties of alloy materials [10]. The present research selected the Ti-8Mo-6Ni alloy and added the different amounts of NbC carbides as a strengthening phase. The effects of the microstructural evolution on the mechanical and corrosion properties of NbC carbides added to Ti-8Mo-6Ni alloys through the vacuum sintering process at various temperatures were determined.

2 Experimental Procedures

In the research, different amounts of NbC powders (1, 3 and 5 wt%) were added to Ti-8Mo-6Ni alloy powders. Meanwhile, the Ti-8Mo-6Ni powder samples with the various added amounts of NbC powder were denoted as NbC-1, NbC-3 and NbC-5, respectively. In this study, the PVA (polyvinyl alcohol) as a binder was added. The green body ($40 \times 6 \times 8$ mm^3) of the powder specimen was produced under a uniaxial pressure at 40 MPa for 300 sec. The vacuum sintering was conducted at 1175, 1225 and 1275°C for 1 h in a 1.3×10^{-3} Pa, respectively. A Microtrac X 100 laser was used to analyze the particle size of the refined Ti, Mo and Ni powders. The mean particle sizes of the reduced Ti, Mo and Ni powders were about 28.4±0.3, 12.4±0.5 and 9.9±0.7 μm, respectively. Moreover, the NbC powder was 6.353 ± 0.215 μm.

To evaluate the effects of NbC strengthening phase on the Ti-8Mo-6Ni alloys *via* vacuum sintering process, this study conducted volume shrinkage rate, relative density, hardness, Transverse Rupture Strength (TRS) tests, X-ray diffraction (XRD) and Scanning Electron Microscopy (SEM) microstructure observations. The Hung Ta universal material test machine (HT-9501A) with a maximum load of 245 kN was used for the TRS tests (ASTM B528-05). Simultaneously, R_{bm} was the TRS, which is determined as the fracture stress in the surface zone. F was maximum fracture load, L was 30 mm, k was chamfer correction factor (normally 1.00-1.02), b and h were 5 mm in the equation $R_{bm} = 3FLk/2bh^2$, respectively. The specimen dimensions of the TRS test were $5 \times 5 \times 40$ mm^3. Porosity test was conducted in accordance to the ASTM C373-88 standard. Hardness tests were measured by HRC with a loading of 1.47 kN, which followed the ASTM E18-3 standard.

Furthermore, corrosion potential analysis uses three electrodes method and follows by ASTM G59-97: the reference electrode is a saturated of silver-silver chloride electrode, auxiliary electrode uses a platinum electrode, and the working electrode is connected to the test specimens. The contact area of the specimen was 0.785 cm^2. The corrosive solvent used 3.5 wt% NaCl was maintained at room temperature. A scanning speed of 0.01 Vs^{-1}, initial potential of -2.0 V, and the final potential of 2.0 V were controlled [7]. The polarization curve was obtained by Corr-View software to analyze and compare the corrosion potential (E_{corr}), corrosion current (I_{corr}) and polarization resistance (R_p) of NbC-1, NbC-3 and NbC-5 sintered specimens.

3 Results and Discussion

Fig. 1 shows the volume shrinkage and relative density of the NbC-1, NbC-3 and NbC-5 specimens through the different sintering temperatures. The volume shrinkage rate of the NbC-1, NbC-3 and NbC-5 specimens significantly increased as the sintering temperature increased, as shown in Fig. 1a. As a result, the NbC-1, NbC-3 and NbC-5 specimens possessed a higher volume shrinkage rate after sintering at 1275°C for 1 h. The highest volume shrinkage rate was 32.9±1.0, 33.1±0.5 and 34.1±0.6%, which appeared in the NbC-1, NbC-3 and NbC-5 specimens after 1275°C sintering for 1 h, respectively. It was reasonable to suggest that the NbC powder added to Ti-8Mo-6Ni alloys effectively filled the voids during the sintering process, which led to the small pores in the specimens decreasing, since the higher temperature sintering enhanced the volume shrinkage rate.

In addition, the NbC-1, NbC-3 and NbC-5 specimens had a higher relative density after sintering at 1275°C for 1 h, as shown in Fig. 1b. The suitable amount of NbC powder in the Ti-8Mo-6Ni specimens advantageous to the relative density resulted from the increase in the diffusion rate at higher sintering temperatures. Through increasing the added amount of NbC powders and the sintering temperature (1275°C), the near-full density of the Ti-8Mo-6Ni alloy could be acquired. Consequently, the highest relative density was 99.7±0.14, 99.6±0.21 and 99.6±0.03%, which appeared in the NbC-1, NbC-3 and NbC-5 specimens after 1275°C sintering for 1 h, respectively. Clearly, the optimal sintering temperature of Ti-8Mo-6Ni alloy adding NbC powders was 1275°C.

Fig 2 reveals the XRD patterns of the NbC-1, NbC-3 and NbC-5 specimens after sintering at 1275°C for 1 h. Previous study indicated that the impurity concentrations of vacuum-sintered commercially pure titanium with fine grains were sufficiently low to meet the ASTM standard. It illustrates that the oxygen, nitrogen and hydrogen contents of the specimens was lower than that of ASTM (B381-10) [2, 11]. As a result, the contamination levels such as oxygen, nitrogen and hydrogen can be ignored. In this work, the XRD patterns also confirm that there are no other oxide appeared in the specimens, as show in Fig. 2. Significantly, α, β titanium and TiC peaks appeared in the XRD patterns of the NbC-1, NbC-3 and NbC-5 specimens after sintering at 1275°C for 1 h. There were no any NbC peaks generated after vacuum sintering process. The experimental results confirmed to our previous study, the oversaturated Ni which precipitated as TiNi intermetallic compounds coexisted with β titanium; moreover, α titanium precipitated plate-like at the grain boundaries of the β matrix. At the same time, the NbC and Ti reaction and transform to TiC after sintering process.

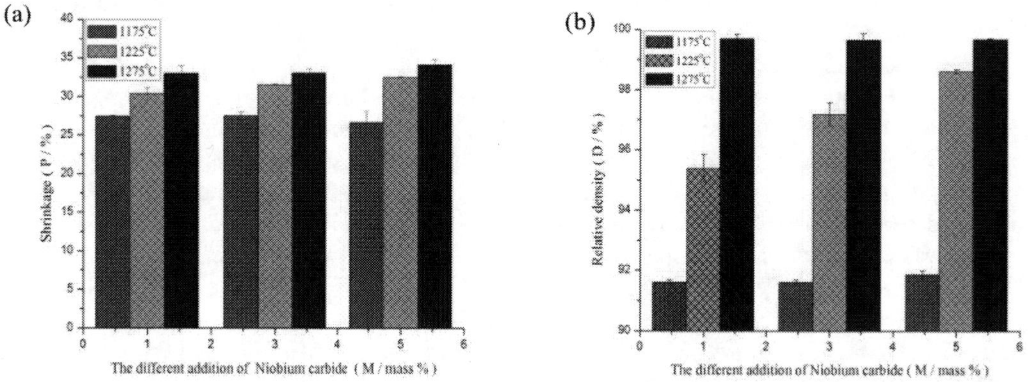

Figure 1. Comparison of the shrinkage and relative density of NbC-1, NbC-3 and NbC-5 specimens after various sintering temperatures

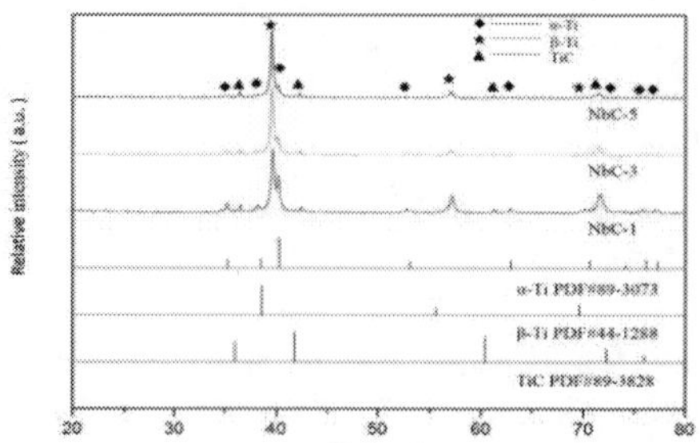

Figure 2. XRD patterns of the NbC-1, NbC-3 and NbC-5 specimens after sintering at 1275°C for 1 h. α titanium precipitated at the grain boundaries of the β matrix, the NbC and Ti react and transform into TiC plus eutectoid β-$Ti(Nb)_x$ phase after sintering process.

DOI: 10.1051/matecconf/201710901006

On the other hand, the SEM morphology observations of NbC-1, NbC-3 and NbC-5 specimens by vacuum sintering at 1275°C for 1 h, as shown in Fig. 3. The porosity significantly decreased after the high-temperature sintering. Moreover, the grain size was obviously decreased as the NbC increased. Among them, the compositions and amounts of the elements of the NbC-1 specimen were determined. A high magnified SEM image is shown in Fig. 4a. Meanwhile, the circular precipitates (black) were TiC (point 1), and EDS analysis as shown in Fig. 4b. The threadlike precipitates (white) were β titanium (point 2) which randomly dispersed within the grains, and EDS analysis as shown in Fig. 4c. Notably, α titanium distributed in the precipitates surrounded. Furthermore, location of the 3 was β titanium (point 3), and EDS analysis as shown in Fig. 4d. These results confirmed that the NbC-1 specimen formed α+β titanium alloys and TiC carbides. The results agree with the XRD finding.

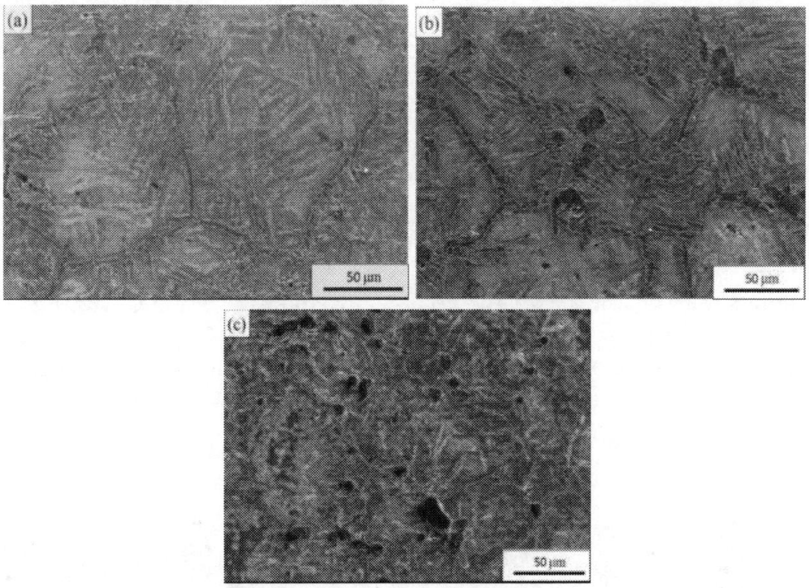

Figure 3. Comparison of SEM morphology observations of (a) NbC-1, (b) NbC-3 and (c) NbC-5 specimens by vacuum sintering at 1275°C for 1 h.

Figure 4. EDS analysis of NbC-1 specimen by vacuum sintering at 1275°C (a) a high magnified SEM image, (b) location 1, (c) location 2, and (d) location 3.

Fig. 5 shows the mechanical properties of NbC-1, NbC-3 and NbC-5 specimens by vacuum sintering at different temperatures. The hardness of the Ti-8Mo-6Ni specimens improved as the sintering temperature increased, and it was dramatically improved as the added amount of NbC powders increased, as shown in Fig. 5a. The previous literature indicates that decreasing the porosities and increasing the density of sintered materials effectively enhances plastic deformation resistance and hardness [2, 12]. In this study, the porosities decreased and the relative density increased as the sintering temperature was raised (1175 → 1225 → 1275°C), which could be the main factor in improving hardness. Besides, the TiNi intermetallic compounds belong to the hard and brittle materials uniformly distributed in the β matrix, which could be another factor in improving the hardness. As a result, the higher hardness appeared in the NbC-1 specimens, and the highest hardness value of HRC 48.1±0.7 appeared after sintering at 1275°C for 1 h. This could be ascribed to the higher relative density (99.7%), dispersion-strengthened TiNi compounds and MC carbides.

Fig. 5b shows the TRS values of the NbC-1, NbC-3 and NbC-5 specimens after sintering at various temperatures. Notably, the TRS values of the Ti-8Mo-6Ni specimens significantly decreased as the added amount of NbC powders increased after sintering at 1275°C for 1 h, and when the added amount of NbC powders increased to 5%, the TRS value obviously decreased (645±36 MPa). The highest TRS value of 976±23 MPa appeared in the NbC-1 specimens after sintering at 1275°C for 1 h. Previous study [2, 13] has indicated that the needle-like microstructure of the Widmanstätten-like structure leads to brittle behavior of the carbon steels. However, it is good for the ductility of the Ti alloys. It is reasonable to speculate that the more Widmanstätten-like structure uniformly distributed in the Ti-8Mo-6Ni specimens after sintering at 1275°C for 1 h. In this study, the amount of Widmanstätten-like structure increase is effective in improving the TRS values. In addition, the excessive precipitation of TiNi intermetallic compounds and MC carbides within the β matrix and the grain coarsening appeared as the added amount of NbC powders increased, which were disadvantageous to the TRS. Consequently, added suitable amount of NbC powders to Ti-8Mo-6Ni alloys was effectively improved the mechanical properties after vacuum sintering process.

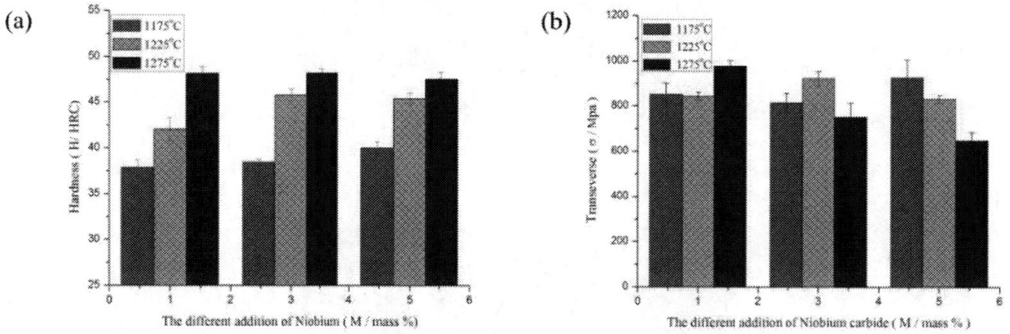

Figure 5. Comparison of the mechanical properties of NbC-1, NbC-3 and NbC-5 specimens by vacuum sintering at different temperatures (a) hardness, and (b) TRS test result.

Fig. 6 shows the Tafel results of NbC-1, NbC-3 and NbC-5 specimens by vacuum sintering at 1275°C for 1 h after 3.5 wt% NaCl corrosion test. All the sintered specimens possessed a significant passivation phenomenon after the 3.5 wt% NaCl corrosion test. Generally speaking, in an electrochemical reaction, the current value represents the diversification of the equilibrium constants in the oxidation reaction. If the current value is higher, it leads to an increase in the equilibrium constant and a fast corrosion rate. However, the polarization resistance must also be considered. Higher polarization resistance usually means better corrosion resistance. In the research, the lowest I_{corr} (3.21×10^{-7} A·cm^{-2}) and highest R_p (29172 Ω·cm^{2}) appeared in the NbC-3 specimens after sintering at 1275°C for 1 h, which effectively improved the corrosion resistance. Furthermore, the NbC-1 specimens also possessed the lower I_{corr} (2.88×10^{-6} A·cm^{-2}) and higher R_p (24768 Ω·cm^{2}) after sintering at 1275°C for 1 h. The result can be further compared with the relative density as shown in

Fig. 1b, it is reasonable to suggest that the electrochemical characteristics are principally controlled by porous features. This could be ascribed to the latter's higher porosity, which resulted from a few pores around the NbC particles. According to the above discussion and results, it was speculated that the NbC-1 specimens sintered at 1275°C for 1 h possessed the optimal hardness and TRS and better corrosion resistance, which result from the more amount and uniform distribution of Widmanstätten-like structure and higher relative density.

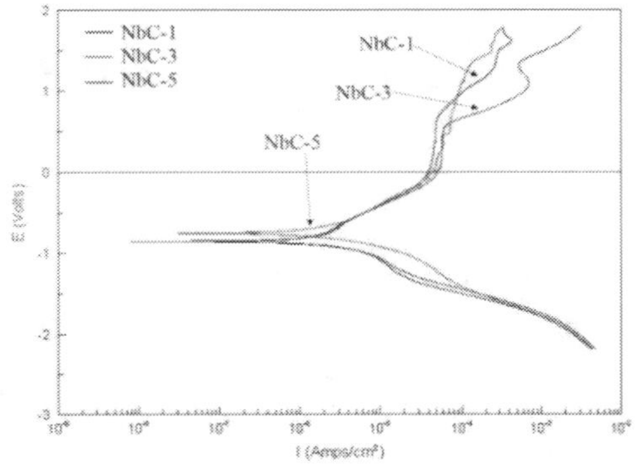

Figure 6. Tafel results of NbC-1, NbC-3 and NbC-5 specimens by vacuum sintering at 1275°C for 1 h after 3.5 wt% NaCl corrosion test.

4 Conclusions

In this work, a highest sintering density (99.7%) and hardness (48.1 HRC) of the NbC-1 specimens were acquired after sintering at 1275°C for 1 h. Meanwhile, the NbC-1 specimens also possessed a highest TRS value (976 MPa). Significantly, the suitably (1 wt% NbC) and uniformly distributed TiNi intermetallic compounds within the β matrix were advantageous to the TRS. On the other hand, the lowest I_{corr} (3.21×10^{-7} $A\cdot cm^{-2}$) and highest R_p (29172 $\Omega\cdot cm^2$) appeared in the NbC-3 specimens after sintering at 1275°C for 1 h, which effectively improved the corrosion resistance.

Acknowledgments

The research is supported by the ASSAB STEELS TAIWAN CO., LTD. The authors would like to express their appreciations for Dr. Harvard Chen and Michael Liao.

References

1. H. Matsumoto, S. Watanabe, S. Hanada, J. Alloy. Compd. **439**, 146 (2007)
2. C. Liang, S.H. Chang, W.D. Wang, K.T. Huang, S.T. Lin, Mater. Trans. **57**, 1363 (2016)
3. W.D. Zhang, Y. Liu, H. Wu, M. Song, T.Y. Zhang, X.D. Lan, T.H. Yao, Mater. Charact. **106**, 302 (2015)
4. S.H. Chang, P.Y. Chang, Mater. Sci. Eng. A. **606**, 150 (2014)
5. F.X. Xie, X.M. He, Y.M. Lv, M.P. Wu, X.B. He, X.H. Qu, Corr. Sci. **95**, 117 (2015)
6. S.H. Chang, J.C. Chen and K.T. Huang, J.K. Chen, Mater. Trans. **54**, 1034 (2013)
7. S.H. Chang, S.L. Chen, J. Alloy. Compd. **585**, 407 (2014)
8. J. Weidow, H.O. Andrén, Int. J. Refract. Met. H. **29**, 38 (2011)
9. T.B. Trung, H. Zuhailawati, Z.A. Ahmad, K.N. Ishihara, J. Alloy. Compd. **552**, 20 (2013)
10. W. Zhou, J. Xiong, W.C. Wan, Z.X. Guo, Z.H. Lin, S.W. Huang, L.M. Tang, H. Zhong, Int. J. Refract. Met. H. **50**, 72 (2015)
11. P. Sun, Z.Z. Fang, M. Koopman, Adv. Eng. Mater. **15**, 1007 (2012)
12. C. Liang, S.H. Chang, J.R. Huang, K.T. Huang, S.T. Lin, Mater. Trans. **56**, 1127 (2015)
13. H. Fujii, Nippon steel technical report. **62**, 74 (1994)

Performance of TiAlN PVD Coated Carbide Drill when Drilling Titanium 6246 Alloy

Mahros Darsin[1, 2,a], Tim Pasang[1] and Zhan Chen[1]

[1]*AUT University, Mechanical Engineering Department, 19 St Paul Street, Auckland, New Zealand*
[2]*University of Jember, Mechanical Engineering Department, Jl Kalimantan 37, Jember, Indonesia*

Abstract. This article describes the TiAlN PVD coated carbide drill tool performance after being used for drilling of Ti-6246 Titanium alloy. Every drill bit was used for single drilling and then being observed with SEM from flank and rake view. Taguchi method L18 design of experiment was used to find the optimum parameters. Wear was not found in this research, however, the tool delamination and tool chipping was evidenced, even if drilling only 10 mm depth. Tool experienced built-up edge in the outer blade, inner blade and at the chisel. This BUE can be seen at both view, i.e. rake and flank sides. BUE is the most dominant deterioration and inevitably in drilling this alloy, regardless the parameters applied. Tool delamination may occur during mechanism of peeling off the BUE. While chipping was related to higher feed rate, that may relate to high MRR. Analysing data using Minitab 17 shows that combination of drilling without cooling, 30 mm deep, cutting speed of 50 m/min and feed rate of 0.15 mm/min would result in the best performance of TiAlN tool for drilling Ti6246 as received.

1 Introduction

One of the common applications of TiAlN is for coating the Tungsten carbide tool. There is two ways of applying TiAlN on the base tool material, by physical vapour deposition (PVD) and chemical vapour deposition (CVD). TiAlN coating has a strong chemical stability, a low thermal conductivity and a high wear resistance at 900°C. The TiAlN coating increases the surface hardness to approximately 3400 – 3600 Hv and improves the resistance to abrasive wear [1]. Other researchers [2, 3] proved that the Al element incorporated in TiAlN coating forms the superficial layer Al_2O_3 to improve the wear resistance and to enhance the chemical stability.

In machining hard metals, PVD TiAlN has shown its better performance than TiN and TiCN [4]. TiAlN also possessed perfect tribological properties compares to TiN, AlTiN and CrAlN coating [5]. TiAlN coated tools have been used for machining hard metals such as Inconnel718 [6], AISI 414 steel [7], AISI 4340 hardened steel [8], Titanium Ti6Al4V [9].

In this research TiAlN PVD coated carbide insert drill is proposed for drilling Titanium alloy 6246. Ti-6246 has some advantages such as corrosion resistance, low heat conductivity, high hardness, non-magnetic and wear resistance. In machining point of view, this alloy among hard-to-machine alloy. There are three main reasons why it so. First, its high hot hardness – retain hardness at high

[a] Corresponding author: mahros.teknik@unej.ac.id

temperature. During machining, heat emerges due to friction between tool and workpiece. The heat generated is not able to soften the alloy. Secondly, it is more severe because this alloy has low heat dissipation. The generated heat was absorbed by the tool (80%), chips (10%) and work material (10%) [10]. That is in contrast with the heat distribution in machining steel the percentage is 10%, 10% and 80% respectively. The last reason is Ti-6246 has high affinity to other materials. In machining, it tends to form built-up edge – debris attachment on the tool.

There are some reasons in the selecting of TiAlN PVD coated carbide for drilling Ti-6246 alloy. TiAlN PVD coated carbide was recommended for higher cutting speed in turning Ti-64 [11]. Ti-64 has some similarities with Ti-6246. In drilling Titanium alloy Ti-64, TiAlN coating tool was more superior to the uncoated tool, it increased tool life by 2416% at the same cutting speed of 25 m/min [12]. In contrast, tool delamination occurred in 5 seconds when face milling of Ti-6246 used TiN coated carbide [13].

To extend of our knowledge, there is no published paper on the performance of TiAlN PVD coated Tungsten carbide for drilling Ti-6246. Studying TiAlN coated drill performance will help to improve machinability of the Ti-6246 alloy.

In this research tool performance was measured by examining the tool condition (or deterioration) after one drilling with certain depth. By definition, tool deterioration is all changes in a cutting part of a tool caused by cutting process [14]. From this definition, all changes of tool may include chipping, wear or addition to the original shape (such as attachment or built-up edge) or crack. Furthermore, ISO define tool wear is a change in shape of the cutting part of the tool from its original shape, resulting from progressive loss of tool material during cutting. Whereas, chipping or brittle fracture is an occurrence of cracks in the cutting part of a tool followed by loss of small fragments of tool material, resulting from crack initiation during cutting. Wear of tool is the condition to measure tool life and tool life itself is one and has been accepted as the most method for machinability judgement criterion [10].

2 Experimental Procedures

The tools used were IC908 Sumocham 10 mm insert type of TAlN PVD coated carbide with included angle 140°. The drill is designed with thinning type – the inner cutting lips has been ground to reduce the chisel side length - the dull part of the tool (Fig 1a). The drill bit then is inserted into the drill body as depicted in Figure 1b. Some blocks of Ti-6246 were prepared with a size of 25 mm x 25 mm and vary in height, from 15 mm to 50 mm. prior to drilling a block was clamped in a fixture which is mounted on a Kistler dynamometer to measure the forces worked during drilling. The dynamometer is connected to a PC outside the CNC machine to monitor the process.

Taguchi L18 with 2 level of single factor and other 4 factors with 3 levels were used for designing the experiments as shown in Table 1.

Table 1. Variation of machining parameters and their level.

Machining Parameters	Levels		
	Low	Medium	High
Coolant	off	-	on
Heat Treatment *)	AR	HT1	HT2
Depth of drilling (mm)	10	30	45
Cutting speed (m/min)	27	35	50
Feed rate (mm/min)	0.08	0.11	0.15

*) Three conditions of material: (1) AR: as received, (2) HT1: Heat treated at 870°C for three hours then followed air cooled (AC) and (3) HT3: Heat treated at 870°C for three hours then followed water quenching (WQ).

Prior to and after drilling, the state of the drill heads were observed using Scanning Electron Microscopy (SEM). To get the constant position observations, a fixture was designed and manufactured from an aluminium block. Two main views of observations of drill heads are shown in Figure 1, i.e. flank view (Fig 1c-i) and rake view (Fig 1c-ii).

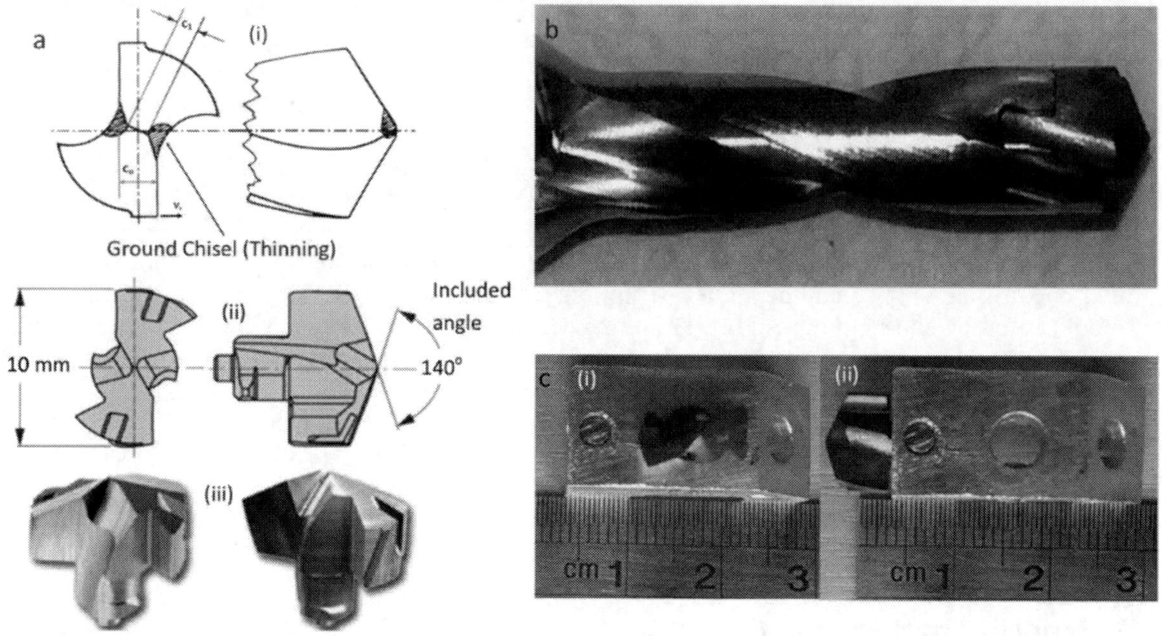

Figure 1. Drill bit type and its main dimension (a), a drill bit mounted on drill body (b), the position of the drill bit when being observed using Scanning Electron Microscopy (SEM): flank view (c-i), rake view (c-ii).

3 Results and Discussions

3.1 Type of Tool Deterioration

Wear was not observed in all drilling condition. It may due to short time or cycles of drilling. Three types of deterioration were observed in this research. Two types were reducing the tool from its original shape: delamination and chipping; and another is an addition to the original tool shape: built-up edge (Figure 2). Built-up edge is an attachment of debris or segments of chips on the tool. According to Tomac [15], there is three category of a built-up edge. It depends on where they stick on. Built-up edge (BUE) stick on the edge of the tool, built-up layer (BUL) on the lip of rake-layer, flank built-up (FBU) on the lip of flank side [15]. However, in this research, we would refer them all as a built-up edge (BUE). Attachment in the form of built-up edge predominated in all condition of drilling. It happened in all parameters used. Usually, BUE still exists in the vicinity of the chisel to the inner blade and less pronounces in the edge of the outer blade. However, the leftover mark on the tool clearly shown that BUE has been delaminated in this site.

Delamination was the second common type deterioration in this research. When the tool has been delaminated the grain of tool base (WC-Co) is obvious as shown in Figure $2b_3$. Delamination is the loss of coating adhesion to a surface or between coating layers.

The last but not the least is chipping. Even though chipping rarely happened, we consider it as the most severe tool deterioration. Chipping is a loss of a small fragment of the tool by sudden action. Consequently, pronounced base tool material can be seen clearly.

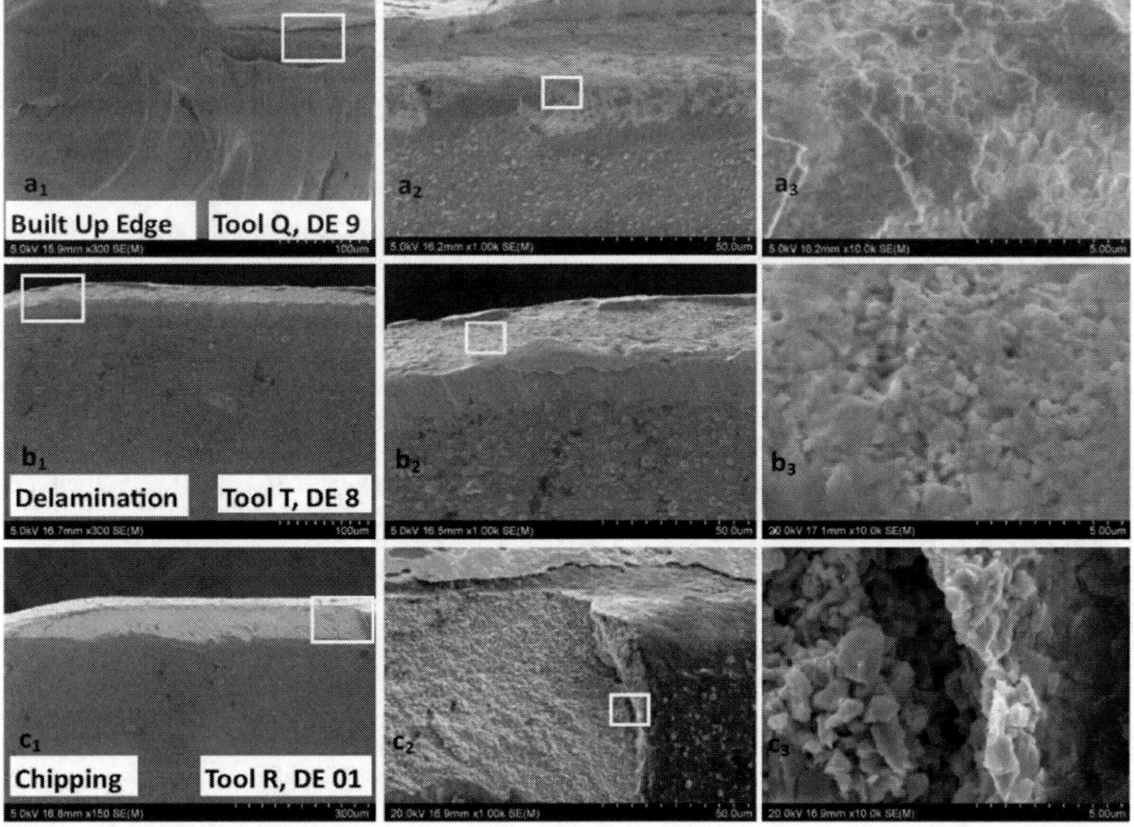

Figure 2. Typical of tool deterioration in drilling Ti-6246.

3.2 Mechanism of Tool Deterioration

Built-up edge is the method of adhering particles present during machining to rake face of cutting tool. During drilling there is some amount of microchips or powdered particles will be produced. According to the first law of thermodynamics that total energy of an isolated system is constant. The energy emerged from the cutting process will be converted into the heat, therefore the rake face is experiencing high temperature. When the microchips or powdered particles are moving the rake face of the tool, they will get exposed to the high-temperature environment and some of the particles may get adhered or welded to the rake face of the tool called as a built-up edge.

Built-up edge formation is very common in machining Titanium alloy due to the affinity of this alloy to most of the tool material [16-17]. Built-up edge does not occur in other material, such as steel, when high-speed machining is applied [18]. Therefore, built-up edge formation in machining Titanium alloys is due to its affinity and low cutting speed applied. Thus, a built-up edge still exists in machining titanium alloy because it is nearly impossible to apply high cutting speed in its machining. In the case of drilling, the formation of BUE is inevitable, even if high speed is applied, because the cutting speed varies from the maximum at the peripheral of the outer blade to zero at the chisel. A thick BUE at the vicinity of the chisel and inner blade is shown in Fig. $2a_1$. Higher magnification (in Fig $2a_2$ and further in Fig $2a_3$) indicates that Carbide particles (the bright part) of base tool material have been revealed.

Delamination of TiAlN coating from the base tool material has also reported by Nouari, M. and Makich, H. (2014) [1]. They concluded that coating delamination was the continuation of adhesion and diffusion modes in machining Ti-64. We reluctance to their opinion that delamination is diffusion mode. Diffusion is a physical process due to different concentration. We could not recognise any different concentration between tool surface and the microchips or powdered particles. We also not in the same opinion to the other researchers [16 & 19] that the delamination is a chemical reaction, to the fact there was no change to a substance at the molecular level. The BUE is formed and then after a while when a critical load is reached the BUE breaks away from the cutting edge. The mechanism of forming and peeling off may repeat thousand times per second [20]. During the peeling off mechanism, some part tool can be torn away. In the case of coated tool, the coating may be delaminated, as seen in Fig $2b_1$, $2b_2$ and $2b_3$ in order of clarity respectively. Actually, the initial delamination has been shown in Fig $2a_3$. Therefore, delamination is a continuation of BUE peeling off.

Fracture or chipping at the edge corner of the drill was reported by previous researchers [16 & 21]. They argued that this chipping process related to BUE peeling off mechanism. According to them, the adherent layer was performed on this site due to high temperature. When the temperature is increased, the adhered layer became weak and could no longer withstand the high compressive stress, then the moving chips pulled out the tool particles along with. In this research, chipping randomly occurs at different places, not only happens at the edge corner of the drill. Fig $2c_1$ shows chipping at the corner between inner and outer blade, about 600 µm long. Figure $2c_2$ shown the edge of chipping area, and from Fig $2c_3$ grains of carbide of base tool material was obvious. Some of the tool chipping associated with high feed rate (0.15 mm/min). Trial on drilling with a feed rate of 0.194 mm/min proved that severe chipping was obvious at the corner between inner and outer blade. It may relate to high metal removal rate (MRR), a volume of the material that has to be removed in a certain time. When a higher volume of material has to be removed in a shorter time, a higher force needed. Therefore, the tool chipping may be caused by a combination of high force and BUE peeling off mechanism.

3.3 Performance of the TiAlN Tool

In regards with the mechanism of tool deterioration, the rank from light to heavy deterioration is BUE, delamination and then chipping. Therefore, we will quantify the degree of deterioration as 1 to 3. The note 1 is for only BUE was observed, note 2 when BUE and delamination have happened, and notation 3 if there was a tool chipping. Observations were carried out at four main section, viz. at the chisel, at the inner blade, at the corner of the inner and outer blade and at the outer blade. Then, total tool deterioration is the cumulative value of four areas of observation. An example of calculation of tool deterioration of design experiment (DE) 1 is presented in Figure 3. On DE1 there are BUE, delamination, chipping and chipping at the four areas of observations. Therefore, the tool deterioration of DE1 is 1+2+3+3=9 (Fig 3a). The result of all experiments is presented in Table 2.

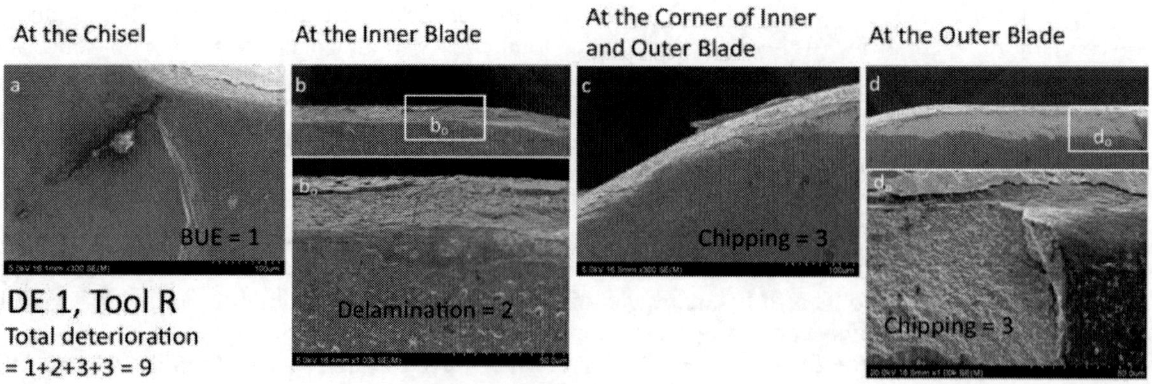

Figure 3. An example of calculating the value of tool deterioration.

Table 2. Tool deterioration as machining parameters changing.

Experiment	Coolant	HT	h (mm)	Vc (m/min)	Fr (mm/rev)	Tool Deterioration
1	No	AR	10	27	0.08	9
2	No	AR	30	35	0.11	6
3	No	AR	45	50	0.15	6
4	No	HT1	10	27	0.11	8
5	No	HT1	30	35	0.15	5
6	No	HT1	45	50	0.08	10
7	No	HT2	10	35	0.08	7
8	No	HT2	30	50	0.11	5
9	No	HT2	45	27	0.15	8
10	Yes	AR	10	50	0.15	4
11	Yes	AR	30	27	0.08	7
12	Yes	AR	45	35	0.11	8
13	Yes	HT1	10	35	0.15	5
14	Yes	HT1	30	50	0.08	7
15	Yes	HT1	45	27	0.11	7
16	Yes	HT2	10	50	0.11	7
17	Yes	HT2	30	27	0.15	10
18	Yes	HT2	45	35	0.08	8

The tool performance is the opposite meaning of the tool deterioration, therefore we were using "Smaller is Better" quality characteristic criteria of main effect plot of means to find minimum effect of tool deterioration of each machining parameters as presented in Figure 4. We can see that variation coolant application gave the least effect on tool performance. To keep the block as received (AR) will give the least effect to deterioration. In our previous publication [22] it was stated that AR most likely to be as forged condition. The HT2 has increased the hardness of the block to 335 HV, in compare to 288 HV in AR condition. HT1 did not change the hardness but it has changed the grain to be smoother than AR [22]. Drilling of up to 30 mm deep result in the lower rate of tool deterioration, while drilling 45 mm lead to rapid tool deterioration. Cutting speed and feed rate influenced greatly to the tool deterioration. Applying either cutting speed of 35 or 50 m/min will not change the tool performance. In term of economics, higher cutting speed is preferable. The higher the feed rate the smaller is the tool deterioration. According to the minimum value of mean value each parameter, it should be an ideal if we apply these following parameters during drilling: with coolant, AR, 30 mm deep, VC 50 m/min and Fr 0.15 mm/min.

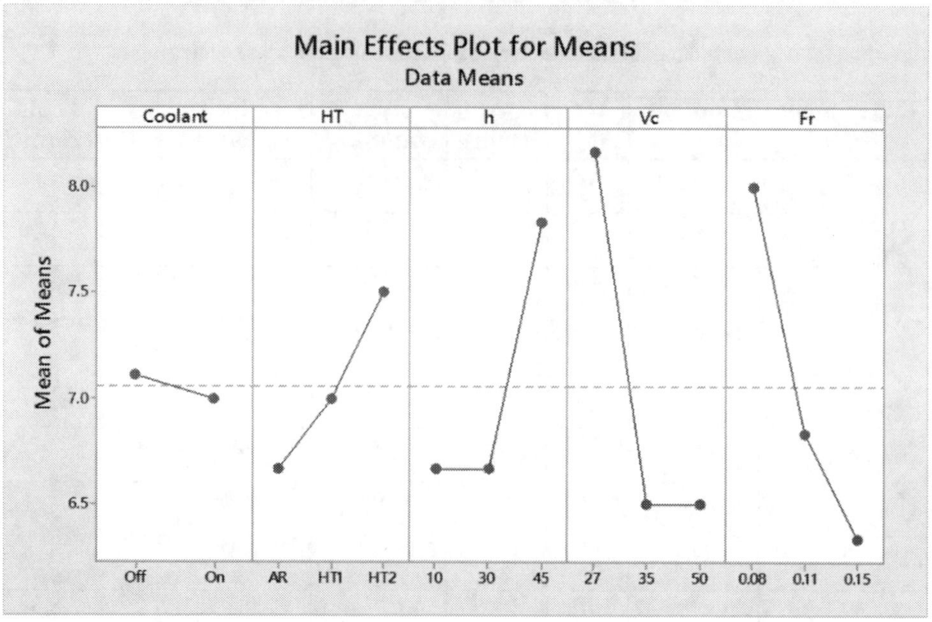

Figure 4. Effect of machining parameters on tool performance.

4 Conclusions

Following the above discussion, we may come to some conclusions. First, the built-up edge is un-avoided tool deterioration in drilling Ti-6246. The cycle of BUE formation through forming and peeling off mechanism may lead to delamination of TiAlN coating from the carbide substrate tool. Whereas, tool chipping was caused by combination of high force and BUE peeling off mechanism. Last, the best performance of TiAlN PVD coated WC-Co carbide drill will be achieved in drilling Ti-6246 block as received (as forged) when using these parameters: with coolant, 30 mm deep, cutting speed of 50 m/min and feed rate of 0.15 mm/min simultaneously.

Acknowledgement

We would like to thank Ministry of Research, Technology and Higher Education of Republic Indonesia through the DIKTI scholarship that financially supports the main author to pursue his PhD degree at the Auckland University of Technology.

References

1. M. Nouari and H. Makich. Metals, **4**, 335-358 (2014).
2. S.S. Gill, R. Singh, H. Singh, J. Singh, Int. J. Mach. Tools Manuf., **51**, 25 – 33 (2011).
3. J. Castanho, M. Vieira, J. Mater. Process. Technol., **143**, 253-257 (2003).
4. P.C. Jindal, A.T. Santhanam, U. Schleinkofer, A.F. Shuster, Int. J. of Refractory Metals and Hard Materials, **17**, 163-170 (1999).
5. L. Aihua, D. Jianxina, C. Haibinga, C. Yangyanga, Z. Juna, Int. J. of Refractory Metals and Hard Materials, **31**, 82-88 (2012).
6. Z. Haoa, D. Gaoa, Y. Fanb, R. Han, Int. J. of Mach. Tools and Manuf., **51**, 12, 973–979 (2011).
7. S.K. Khrais, Y.J. Lin, Wear, **262**, Issues 1–2, 64–69 (2007).
8. CH.R.V. Kumar, P. K. Nair, B. Ramamoorthy, Int. J. Machining and Machinability of Materials, **3**, Nos. 1/2 (2008).
9. N. Andriya, P.V. Rao, and S. Ghosh, *Proc. of the WCE* **III**, London, U.K., 2012).

10. S. Pervaiz, I. Deiab, B. Darras, A. Rashid, M. Nicolescu, Adv. Mat. Research, **685**, 68-75 (2013)
11. S. Pervaiz, A. Rashid, I. Deiab, & M. Nicolescu, Mat. and Manuf. Proc., **29** (3) 219–252 (2014).
12. S. Sharif & EA. Rahim, J. Mat. Proc. Tech., **185** (1-3), 72–76. (2007).
13. S. Sharif, E.A. Rahim, A.S. Mohruni, & A. Jawaid, *Int. Conf. on Leading Edge Manufacturing in 21st Century*. Nagoya (2005).
14. *ISO 8688-2*, International Organisation for Standardization, Genève (1989).
15. N. Tomac, K. Tonnessen, F.O. Rasch, & T. Mikac, *Advanced Manufacturing Systems and Technology*, CISM Courses and Lectures No 486, Springer Wien, New York (2005).
16. J.L. Cantero, M.M. Tardío, M. M., J.a. Canteli, M. Marcos, M., & M.H. Miguélez, Int. J. of Mach. Tools and Manuf., **45** (11), 1246–1255 (2005).
17. E.A. Rahim, S. Sharif, Z. Ahmad, A.S. Mohruni, & I.A. Syed, *Int. Conf. on Leading Edge Manufacturing in 21st Century* (pp. 3–7). Nagoya, (2005).
18. J. Kümmel, J. Gibmeier, E. Müller, R. Schneider, V. Schulze, & A. Wanner, Wear, **311** (1-2), 21–30 (2014).
19. M. Nouari, A. Ginting, Surface & Coating Tech. **200**, 5663 – 5676, (2006).
20. M.A. Dittrich, in answering the question in Research date "How does the built-up edge lead to surface damage" https://www.researchgate.net/post/How_does_the_built-up_edge_lead_to_surface_damage [October 1, 2016]
21. Z. Rihova, K. Saksl, C. Siemers, & D. Ostroushko, WASET, **68**, 1515–1518 (2012).
22. M. Darsin, T. Pasang, Z. Chen, Twenty-fifth Int. Conf. on Processing and Fabrication of Advanced Materials (University of Auckland, New Zealand, January 2017)

Preparation and Sintering Behaviour of Alumina Powder by Ammonia Precipitation Method

Liuyan Wang[1, 2,a]

[1] *School of Architecture and civil engineering, Shenyang University,110044, China*
[2] *Key Laboratory for advanced materials of Liaoning province, Shenyang University, 110044, China*

Abstract. In this paper, alumina precursor was prepared by the ammonia precipitation method which used Al $(NO_3)_3 9H_2O$ as aluminum source and NH_4OH as a precipitator, adding a small amount of PEG4000 as the surface active agent. Finally γ-Al_2O_3 was obtained at 900°C for 2h. The stable alumina crystal form of α-Al_2O_3 was got at 1100°C for 2h. The influence of precipitation agent on the precursor was studied by means of TG / DTA and Tem, XRD etc. The effects of the synthesis temperature and time on the phase composition and morphology of the alumina powder were also analysed.

1 Introduction

Alumina powder as one of the special functional materials, due to the high purity, fine particles and uniformly distributed, it has a broad prospect of application on the ceramic structure, catalyst carrier, fine ceramics for its small size , surface and quantum effect[1].

Preparation of nano alumina powder by precipitation method [2,3,4] has the advantages of low cost, high purity, large amount of production, simple equipment manufacturing and short process. At present, the main precipitation method for preparing alumina powder was used ammonia and urea as precipitation agent,by the reaction of aluminum ions and precipitation agent directly generated γ-AlO(OH),and then sintered this alumina precursor[5].

Based on the preliminary work [6,7], the process parameters such as thermal analysis, synthesis temperature and reaction time were studied by using ammonia precipitation method which took aluminum nitrate as raw material. This method had the advantages of simple process, low cost and short process flow, and it was suitable for the preparation of nanometer powder in the process of industrial production.

2 Experiments

2.1 Raw materials

Raw materials used in this experiment was shown in Table.1.

[a] Corresponding author: w.ly428@163.com

Table.1 Raw materials of experiments

Reagent name	Molecular formula	Specifications	Manufacturer
ammonia	$NH_3 \cdot H_2O$	A.R	Shenyang Chemical Reagent Factory.
Polyvinyl alcohol	PEG4000	A.R	Tianjin kermel chemical reagent
Aluminum nitrate	$Al(NO_3)_3 \cdot 9H_2O$	A.R	Shenyang Chemical Reagent Factory
Absolute alcohol	C_2H_5OH	A.R	Shenyang Chemical Reagent Factory

2.2 Experiment method

Using $Al(NO_3)_3 \cdot 9H_2O$ as aluminum source, NH_4OH as precipitator, a small amount of PEG4000 was added in the experiment as the surface active agent to prevent the aggregation of the product particles.In order to reduce growth rate of the nucleation , the experimental temperature should be controlled below 10 ℃, dropping speed was controlled at 3 ~ 5 ml/min, and the end point pH value was kept between 7 and 8.

After the end of the titration, the preparation of the precursor solution was prepared by aging and centrifugation, and then washed with anhydrous ethanol.

Finally, the resulting sample was placed in a rapid heating furnace, with a temperature of 10℃/min rise to 900℃ for 2h, the sample was token with the furnace cooling, and finally ultra-fine alumina powder was got.

3 Results and discussion

3.1 Effect of dispersant on particle size and morphology

Surface active agent as a good chemical dispersants can be adsorbed on the surface of the powder particles. It made the inter particles difficult to reunite and made the powder dispersed by electrostatic repulsion and VDW[8,9,10]. Some research results shown that PEG4000 can form a uniform and thick protective film on the surface of the colloid, which can restrain the aggregation of colloidal particles very effectively[11]. Therefore, PEG4000 was used as a dispersant to prepare ultrafine alumina powders.

3.2 TG-DSC comprehensive thermal analysis of alumina precursor

The TG-DSC curve of alumina precursor prepared by ammonia as precipitator was shown in Figure 1. With the increase of sintering temperature, weightlessness of $Al(OH)_3$ was a continuous process when the temperature was below 700℃. The curve showed a continuous decline in this process. When the temperature was higher than 700℃, the TG curve was almost a horizontal curve. This horizontal curve indicated that the decomposition of $Al(OH)_3$ was over. The TG curve was in a horizontal state

DOI: 10.1051/matecconf/201710902002

for the $Al(OH)_3$ has begun to phase transformation,and the phase change process was a constant weight process.

An obvious endothermic peak was appeared in the DSC curve at 133.46℃.The adsorption water was divided from the sample. Endothermic peaks was appeared nearly 207.80℃ for the volatilization of residual organic matter in the sample. A strong endothermic peak appeared at 245.98℃ for the alumina precursor was appeared the following decomposition reaction, see formula (1):

$$Al(OH)_3 \rightarrow AlOOH + H_2O \quad (1)$$

Water produced by the decomposition reaction was overflowed in the form of water vapor. With the increase of temperature, a strong exothermic peak was appeared at 299.77℃ for AlOOH was generated from the decomposition reaction, see formula (2). TG curves between 200℃ to 300℃ sharp declined for AlOOH was changed to Al_2O_3 and Decomposition of water was evaporated from the sample.

$$2AlOOH = Al_2O_3 + H_2O \quad (2)$$

When the heating temperature rises to 1107℃ and 1163℃ respectively, two obvious endothermic peak were correspondingly appeared on the DSC curve. There was no obvious phenomenon of weightlessness in the TG curve. Stable alumina crystal can be obtained at 1000℃.

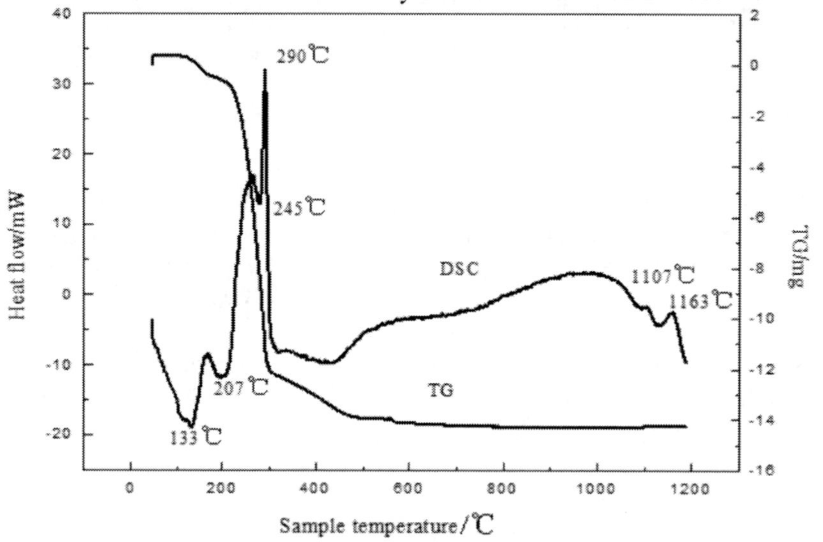

Figure 1. TG-DSC curve of Al_2O_3 precursor prepared by ammonium hydroxide

3.3 Effect of calcination temperature on the phase composition of alumina

The XRD diagram of alumina precursor prepared by ammonia water as the precipitation agent at different temperatures was shown in Figure 2. Amorphous Al_2O_3 was obtained when the sintering temperature reached 800℃ for 2h as shown from Fig.2[12].

γ-Al_2O_3 was got when the temperature rises to 900℃ for 2h. γ-Al_2O_3,δ-Al_2O_3,α-Al_2O_3 and other types of crystal were appeared at 1000℃. The precursor was almost completely converted toα-Al_2O_3 at 1100℃. This phenomenon showed that with temperature increased, the composition of the structure changed obviously. And the stability of the structure was increased. Finally, the structure of α-Al_2O_3 with stable crystal structure was obtained.

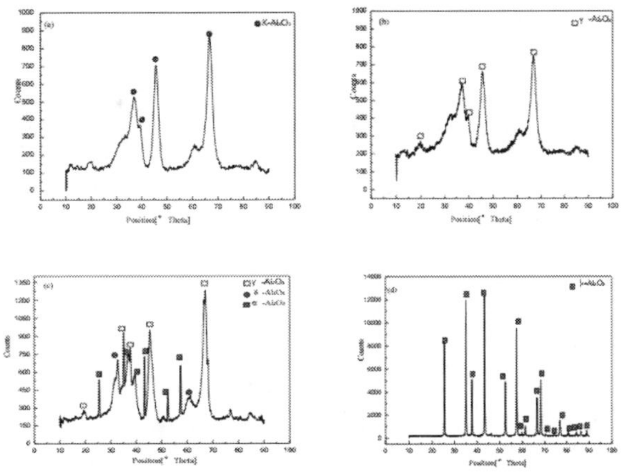

Figure 2. XRD pattern at different calcination temperature preparedby ammonium hydroxide

3.4 Effect of sintering time on the weight loss rate of alumina powder

The sintering temperature range was determined by the comprehensive thermal analysis of alumina precursor. In order to determine the reasonable holding time, a certain amount of alumina precursor was taken under 900℃and incubated for 1 hour, 1.5 hours, 2 and 2.5 hours. The relationship between the powder weight loss rate and the sintering time was as shown below in Fig.3. The weight loss rate of the powder was basically unchanged at 900℃ for 2 hours compared with followed sintering time. Therefore, 2 hours of the sintering time was chosen due to the aggregation and growth of grains can be avoided.

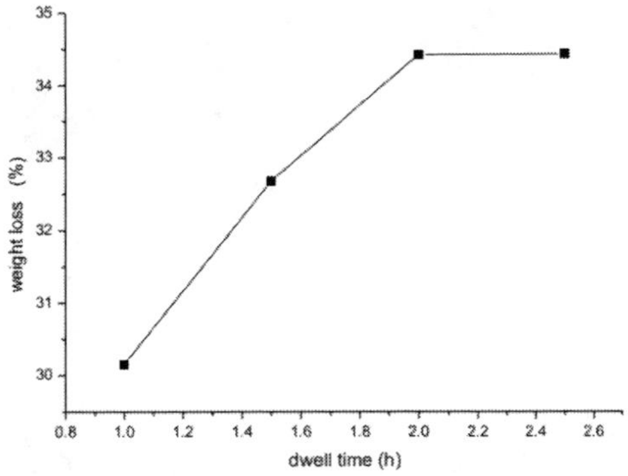

Figure 3. Relation between weight loss and dwell time

3.5 Effect of calcination temperature on particle size and morphology of alumina powder

The relationship between different calcinationtemperature and average particle size of alumina powder prepared by using ammonia as precipitator was shown in Fig. 4.

The different sintering temperature had a significant effect on the particle size of alumina. At the same sintering time, with the increase of the sintering temperature, the relationship between the sintering temperature and the average particle size firstly decreased, then increased followed by decreasing. And the minimum size of the particle size was obtained at 900℃.

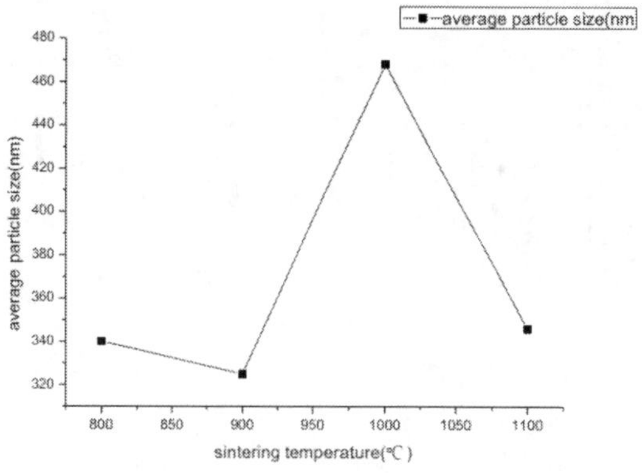

Figure 4. Relationship between different sintering temperature and average particle size with ammonia as precipitator

The nano scale spherical alumina powder can be obtained from alumina precursor sintered by 900 ℃for 2 hours, as shown in Fig.5.

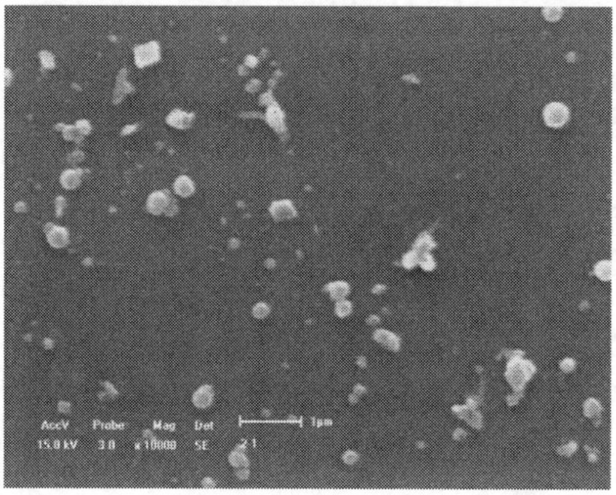

Figure 5. SEM image of Al_2O_3 powder prepared by ammonium

4 Conclusions

Nano alumina particles was got from the dried precursor by sintering 900℃ after 2 hours. Alumina precursor was prepared which used 0.3mol/l aluminum nitrate as raw material and 1:2 ammonia as precipitator, adding a certain amount of dispersant. In this precipitation process, temperature was controlled below 10 degrees. And pH was between 7 and 8.

After adding the dispersant PEG4000, fine particle size, appearance and good dispersion of the alumina powder can be obtained. In the sintering process of alumina precursor, the crystal components increased gradually, and the crystal composition increased with the increase of the sintering temperature. A stable crystal structure of the -Al_2O_3 can be obtained with ammonia as the precipitation agent prepared by the alumina precursor sintered at 1100℃ for 2 h.

References

1. T.Zhu,F.M.Wang,X.D.Wang,etc. Progress and application of nanometer materials in foreign countries. Chemical Industry Press,2002.

2. Y.J.Wang, C.X.Lee, Z.H.Wang. Preparation of nano alumina particles by ultrasonic chemical precipitation method. Journal of Beijing chemical industry, 29(2002).
3. Y.F.Chang, G.L.Ning, Y.Lin. Application of a new type of temperature resistant surface active agent in the preparation of spherical alumina nano powder. Powder technology in China, 6(2000).
4. I.N.Bhattacharya, P.K.Gochhayat, P.S.Mukherjee, etal. Thermal decomposition of precipitated low bulk density basic aluminium sulfate.Materials Chemistry and Physis, 88(2004).
5. A.F.Zhang,J.P.Liu. Study on the new technology of preparing nano alumina powder by precipitation method, Inorganic Chemicals industry, 35(2003).
6. Y.Q.Wu, Y.F.Zhang, X.X.Huang. Preparation of nanoα-Al_2O_3 powders at low temperature, Journal of inorganic materials.16(2001).
7. P.Wang, D.Z.Wei. Kinetic Analysis on Preparation of Micro Power Aluminum Hydroxide by Liquid Precipitation Process, Journal of Northeastern University(Natural Science), 35(2014).
8. X.J.Sun. Study on the Application of surfactants in preparation and dispersion of Ultra-fine Alumina, Powder Metallurgy Technology.33(2015).
9. G.H.Liu, Y.M.Zhang, X.B.Lee. Agglomeration Mechanism of surfactant of ultra-fine $Al(OH)_3$in drying process, The Chinese Journal of Nonferrous Metals, 19(2009).
10. S.Y.Lu, Y.Q.He. Application of sursurfactant to synthesis of nano-alumina, Materials review, 5(2006).
11. F.GU, Y.Shen, C.Xu, etc. The influence of polymerization degree of dispersing agent on the power properties of nano alumina, Functional material, 36(2005).
12. L.D. Shang. Preparation and identification of eight kinds of crystal alumina, Chemical world, 7(1994).

Effects of Friction Stir Welding on Corrosion Behaviors of AA2024-T4 Aluminum Alloy

Steve Korakan Ales[1], Lei Wang[2a]

[1]*Department of Mechanical Engineering, Auckland University of Technology, Auckland, NZ*
[2]*Department of Electromechanical Engineering, Shenyang Aerospace University, Shenyang, China*

Abstract. In this work, the corrosion behavior of welded joints of AA2024-T4 Al alloy produced by friction stir welding process has been investigated. Tests were performed in an aerated 3.5% NaCl aqueous solution with pH = 7 at 20±2°C. Corrosion rate and corrosion morphology of weld regions were evaluated and compared to those of the parent metal. The microstructure of weld nugget, thermomechanical affected zone, heated affected zone, and parent metal were analyzed using scanning electron microscopy and energy dispersive spectroscopy. It was observed that corrosion initiated at FSW related spots and the sizes of local corrosion increased with time.

1 Introduction

Friction stir welding (FSW) is a solid state joining process patented in 1991 by The Welding Institute (TWI). Being a solid-state process, FSW can be used to weld almost any type of metal, including some previously un-weldable precipitation strengthened 2xxx and 7xxxs Al alloys [1]. Under tensile loading, failure of FS welds most often occur in the area between the heat affected zone (HAZ) and the thermomechanically affected zone (TMAZ), where high temperatures dissolve and coarsen the strengthening precipitates, creating local strength minima during FS [2]. Due to the change in precipitate morphology, this region is also generally susceptible to corrosion [3]. The goal of this research was to identify the corrosion mechanisms on FS Al 2024-T4 Al alloys and on the fatigue behaviors.

Aircraft structures made from high-strength Al alloys exhibit a strong propensity to pitting [4]. Corrosion-fatigue in Al alloys generally involves the formation of pits, their growth, nucleation of cracks from pits, and the eventual propagation of the crack leading to failure. Although pits can initiate from both physical and chemical heterogeneities on the surface, the role of inclusions and second-phase particles in inducing pitting corrosion in aluminum alloys is more common. As the alloys contain numerous such particles, electrochemical reactions are triggered between them and the surrounding matrix [5]. Particle-induced pitting corrosion in aluminum alloys has indeed been observed by scanning and transmission electron microscopic techniques [4]. Pitting corrosion is therefore an issue that continues to cause serious concern in the aircraft industry [6].

[a] Corresponding author: leiwng@sau.edu.cn

DOI: 10.1051/matecconf/201710902003

The mechanism of pitting corrosion and its impact on fatigue in aircraft aluminum alloys has been widely discussed [7]. Corrosion pits are seen to significantly shorten the fatigue crack initiation, decrease the threshold stress intensity by 50% or more, and lower the fatigue strength by about 40% [8]. The pit growth that causes the fatigue crack to initiate is seen to be controlled by the limiting cathodic current density supported by the exposed constituent particles within a growing pit [9].

2 Experimental procedures

There were three steps in conducting the experiments. The first step was to carry out FSW. This was done using a FSW 3LM-4012 machine that uses a tapered needle right hand thread rotation. The diameter of tool was 15 mm, pin diameter was 4.5 mm, length of the pin was 3.7 mm. Tool was made using H13 tool steel. Tilt angle of the tool during FSW was 2°. Tool rotation speed was 1000 rpm and the welding speed was 100 mm/min. The test material used was a 40 mm × 25 mm × 4 mm thick aerospace 2024-T4 Al alloy plate. The chemical composition by mass fraction and mechanical properties are shown in Table 1 and Table 2, respectively.

Table 1 Chemical Composition of Al2024-T4 (wt%)

Al	Cu	Mg	Mn	Si	Fe	Others
91.1	4.4	1.5	1.5	0.5	0.5	0.5

Table 2 Mechanical properties of Al2024-T4 Aluminum Alloy

Tensile strength, MPa	0.2% proof strength, MPa	Elongation, %
420	260	15

Weld burrs along the seam are removed before cutting the samples into 40mm x 25mm x 4mm for testing according to ASTM G31-72 standard for Immersion Corrosion Test for welds. Samples were then submerged in deionized water in a 3.5% NaCl solution (simulated marine environment). The pH value was at 7 and tests were conducted at room temperature (20±2°C). Immersion times were 24, 48, 72, 120, 240 and 480 hours respectively. For 240 and 480 hour immersion, the etchant was replaced after every seven days. For each set of test, two parallel specimens were used for each time. After each test, corrosion products were removed using dilute nitric acid and cleaned using an ultrasonic vibration cleaning equipment. This was followed by static measurement of weight loss due to corrosion rate as defined in equation (1) using a simple weighing balance. Static weight loss method was used to weigh the weight loss of the specimens after corrosion. Equation (1) in the following is used to calculate the corrosion rate (CR):

$$\mathbf{CR}\left(\frac{g}{m^2.h}\right) = 1.0x10^4\ x\ W/(A\ x\ T) \tag{1}$$

where W is the weight loss, A is corrosion area, T is corrosion time.

Note that the corrosion rate of the weld is calculated herein as the average corrosion rate, and therefore does not reflect the extent of corrosion of a specific specimen. During a test, corrosion normally started by pitting and the number of pits increased fast at the beginning, while a higher rate and more serious form of corrosion developed later. It is necessary to combine the mass loss and the corrosion rate to comprehensively analyze the severity of the specimen under consideration. Therefore, equation (2) defines the percentage material removal as follows:

$$W_L = \left(\frac{(W_B - W_A)}{W_B}\right) x\ 100\% \tag{2}$$

where, W_L is the % weight loss ratio, W_B is the weight before corrosion, W_A is the weight after corrosion

Hardness tests of the FS welds were carried out by using Vickers hardness instrument. Micro-hardness tests are done with an HVS-1000A Vickers hardness tester across the weld section at evenly spacing of 0.25mm near the TMAZ and 0.5mm for the other zones. The weld was observed using Olympus GX51 optical microscope after corrosion. Characteristics of corroded pits macro-morphology photos were taken for each time per sample using an FEI-QUANTA 600 scanning electron microscope (SEM) with its digital camera, combined with EDS analysis of the corrosion products and phases.

3 Results and analysis

3.1 Microhardness and microstructure

A hardness profile is shown in Fig. 1. Clearly, hardness varies significantly at different points from the centre of the weld. The lowest hardness, approximately 92.2 HV, is found in the TMAZ region approximately 4 mm from the plate joint line. Hardness increases toward the Nugget slightly. A higher hardness, approximately 128.8 HV, is found in the nugget. Hardness in the parent material is much higher than the weld zones reaching around 130 HV. The softening in the weld zone is due to annealing in the zone during FSW.

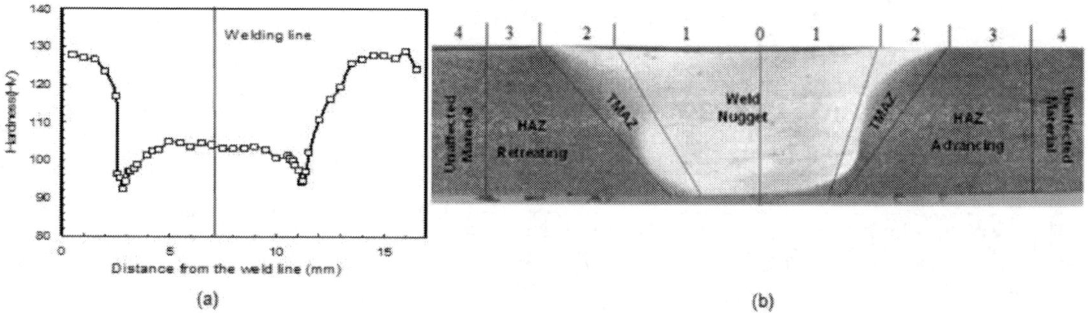

Figure. 1 (a) Vickers hardness transverse across the friction stir welded 2024-T4 aluminum alloy joint at various positions in the section. (b) Different FS Weld Zones affected by weld too during FS welding.

Microstructure in each region of a FSW joint is shown in Fig. 2. The finest microstructure is found in the nugget zone (NZ) which has an equiaxed grains of the order of 5 to 10 μm (Fig.2b). Microstructure present in the TMAZ shows an elongated grain structure that is believed to be due to severe plastic deformation that has taken place during welding, because the stirring action causes the grains of the parent metal to be drawn into and around the nugget zone (Fig.2c). The grain structures in the HAZ and parent plate are similar, with a size of the order of 150 to 200 μm (Fig. 2a and 2d), but we can find more second phase precipitates from the matrix because of the heat from FSW. The variation in grain size in each region depends on the material and process parameters of FSW. The non-uniform microstructure across the FSW joints is believed to introduce different fatigue strength for each zone of the weld.

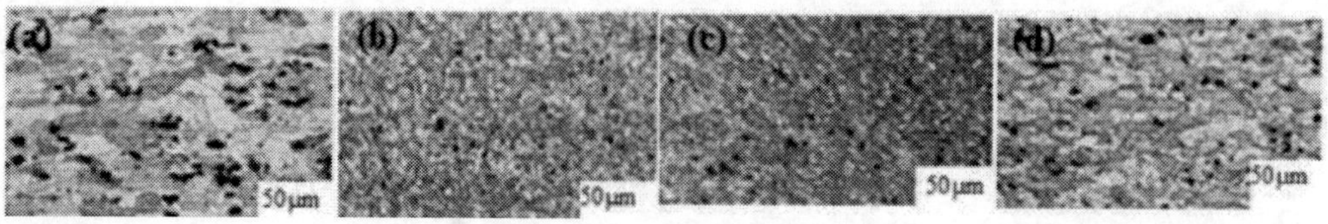

Figure. 2 Microstructures of (a) parent material, (b) nugget (N) region, (c) TMAZ and (d) HAZ.

3.2 Corrosion test results

The following factors may affect the stability of Al oxide and thereby cause corrosion:

- The oxide is not stable in acidic (pH < 4) or alkaline (pH > 9) environments [9]
- Aggressive ions (chlorides, fluorides) may attack the oxide locally.
- Certain elements (Ga, Ti, In, Sn, Pb) may become incorporated in the oxide and destabilize it [10].

Pitting is a highly localized type of corrosion in the presence of aggressive chloride ions. Pits are initiated at weak sites in the oxide by chloride attack [10, 11]. Pits propagate according to the reactions while hydrogen evolution and oxygen reduction are the important reduction processes at the intermetallic cathodes.

As a pit propagates the environment inside the pit (anode) changes. As Al dissolves following reaction (3), according to reaction (4), pH will decrease. To balance the positive charge produced by reaction (3) and (4), chloride ions will migrate into the pit. The resulting HCl formation inside the pit causes accelerated pit propagation. The reduction reaction will cause local alkalinisation around cathodic particles. Al oxide is not stable in such environment, and Al around the particles will dissolve (alkaline pits). The active Al component of the particles will also dissolve selectively, thereby enriching the particle surface with Fe and increasing its cathodic activity. Etching of the Al matrix around the particles may detach the particles from the surface, which may repassivate the alkaline pits. This may also reduce the driving force for the acidic pits causing repassivation of some in the long run.

$$\mathrm{Al} = Al3+ + 3e- \tag{3}$$

$$Al_3^+ + 3H_2O = \mathrm{Al\,(OH)_3 + 3H^+} \tag{4}$$

$$2H+ + 2e- = \mathrm{H_2} \tag{5}$$

$$O_2 + 2H_2O + 4e- = \mathrm{4OH^-} \tag{6}$$

Initial corrosion or pitting corrosion mechanism is based on the second phase dissolution and it started from the weld surface as can be seen in Fig. 3 (a-d) with many small pores and shallow pits. 2024-T4 Al alloy has S phase (Al_2CuMg) and theta phase (Al_2Cu) as the main strengthening phases. Since corrosion potential and matrix phase are similar so that the initial galvanic pitting are established between the S phase and the Al alloy matrix. S phase is about 2.7% of the surface area of the alloy due to its lower corrosion potential, and the preferential dissolution of the anode. When the S phase of the Al, Mg element after dissolution, the remaining elements with a high Cu is electrochemically inert to a large number of enrichment.

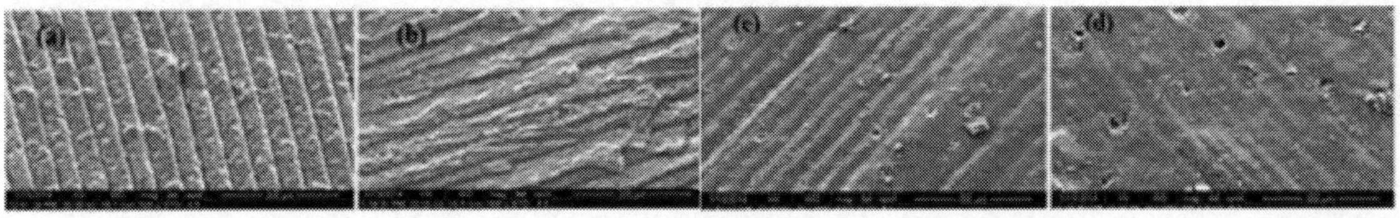

Figure. 3 Pits Morphologies of Weld Surfaces (a) TMAZ, (b) HAZ advancing side, (c) HAZ Retreating side and (d) WN.

Fig. 4 (a, b) shows the surface of the weld at this copper-red color. Enrichment of Cu led to S phase corrosion potential rising above the Al alloy matrix when potential polarity reversal would occur. S phase acted as the cathode and the anode was the dissolving phase meaning corrosion pits, as can be observed in Fig. 4 (a, b, c, d). At this point, Cu element enrichment was the cathode where oxidation reaction took place and the weld surface gradually darkened as shown in Fig. 4 (c, d, and e).

 DOI: 10.1051/matecconf/201710902003

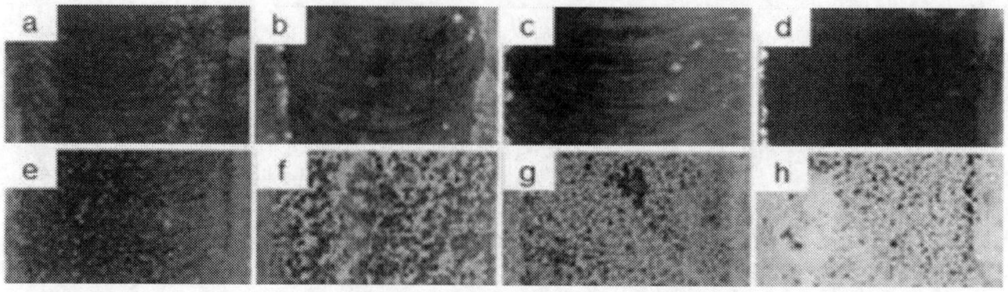

Figure. 4 Macro Morphologies of Corroded FS Weld Surfaces. (a) 24hr (b) 48hr (c) 72hr (d) 96hr (e) 120hr (f) 240hr (g) 360hr (h) 480hr.

With the further development of the corrosion process, the corrosion product began to appear white granular Fig. 4 (e), identified as mainly Al $(OH)_3$ and Al_2O_3, containing a small amount of Mg oxide. When the immersion time was nearly 240 hours, white corrosion particles increased, and further immersion into 480 hours, the weld surface has been covered by a white corrosion product layer, as show in Fig. 4 (f, g, h). Fig. 5 shows pitting at different set of times. The longer the immersion time, the deeper the pit depth. It is obvious that the depth of the pith depends on corrosion time.

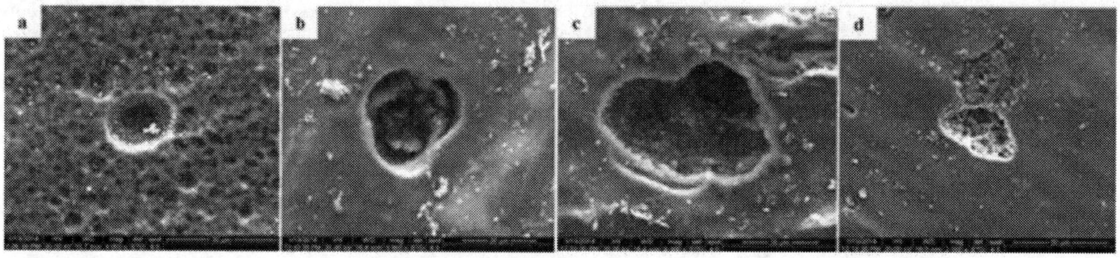

Figure. 5 Morphology of Single Pits. (a) 48h ×4000 (b) 96h ×4000(c) 240h ×4000(d) 480h ×2000.

Decline in Cl^- at the corrosion pit enrichment of pH further promote anodic dissolution in the formation of occluded cell autocatalytic effect. Meanwhile, due to the strengthening phase, mainly in the grain boundaries which appear weak around the Cu precipitation area while there will be a small amount of $MnAl_6$ dispersed in the grain boundaries. This will make the composition of the multiphase system intergranular in the corrosive medium containing Cl^- ions resulting in the occurrence of intergranular corrosion along the grain boundary forming anodic dissolution channels. This stage is mainly due to excessive pitting to intergranular corrosion, where the test piece has less mass loss. Thus corrosion rate is decreased. The spectrum analysis of the weld surface of the elements present in the Cu corrosion after 48 and 480 hours as shown in Fig. 6.

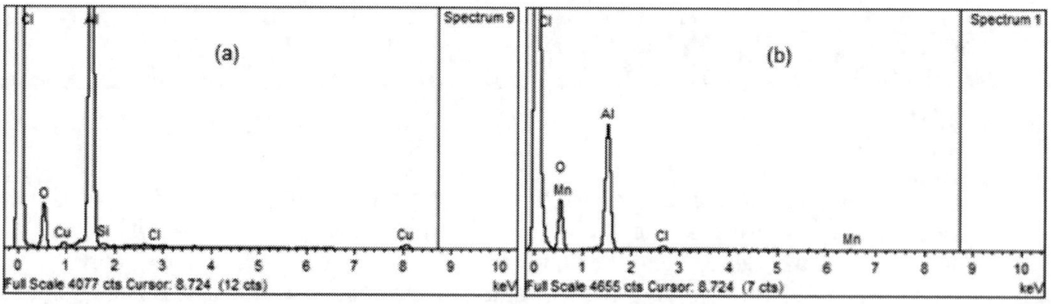

Figure. 6 Results of EDS Analysis (a) Corrosion 48 hr, (b) Corrosion 480 hr,

When immersion proceeded to 480 hours, the development of multiple weld surface corrosion occurred resulting in increased corrosion rate weight loss as shown in Fig. 7 (a, b). Thus, from the initial weld pitting corrosion develops further denudation, erosion leading to more serious specimen mass loss. Therefore, the corrosion occurs at faster rate.

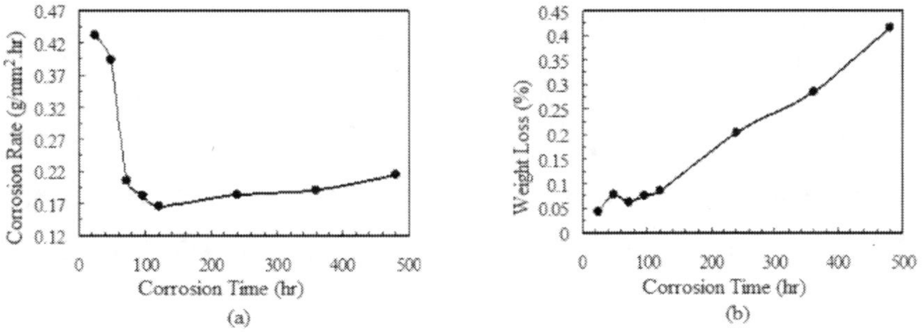

Figure. 7 (a) Corrosion Rate, (b) % weight loss.

4 Conclusions

In age-hardened AA2024-T4 FS weld of thin sheets hardness variations are bigger and are of W-shape with local maximum at weld centre. Variation of hardness (HV) of the FS weld is very lower at the TMAZ region, nugget, HAZ and higher at the parent material.

In FS weld of 2024-T4 aluminium alloy experiences a slow recovery process between the initial transition pitting mechanism and intergranular corrosion.

Due to uneven weld microstructures and precipitate distribution most serious corrosion occurs at weld nugget area, forward side and lighter at the return side.

Pitting in Al-Cu alloys are due to Cu enrichment and redistribution where oxide film breakdown during FS weld which supports anodic reaction.

The longer the immersion time, the more serious is the corrosion of the specimen where corrosion rate increases ultimately due to denudation.

References

1. W.M. Thomas et al. "Friction Stir Welding," U.S. Patent No. **5,460,317**, October, 24, 1995.
2. Genevois, C. et al., *Acta Materialia*, vol. **53**, no. 8, May, (2005), pp. 2447-2458.
3. Hannour, F. et al. *Proceedings of the 2nd International Friction Stir Welding Symposium*, 26-28 June, 2000.
4. P. L. Threadgill: *Sci. Technol. Weld. Join.*, (2007), **12**, 357-360.
5. A. P. Reynolds, W. D. Lockwood and T. U. Seidel: *Mater. Sci. Forum*, 2000, 331-337, 1719-1724.
6. L. E. Murr, G. Liu and J. C. McClure: *J. Mater. Sci.*, (1998), **33**, 1243-1251.
7. K. A. A. Hassan, P. B. Prangnell, A. F. Norman, D. A. Price and S. W. Williams: *Sci. Technol. Weld. Join.*, 2003, 8, 257-268.
8. Y. K. Yang, H. Dong and S. Kou: *Weld. J.*, 2008, **87**, 202s-211s
9. Davis, J. R. et. al, editor, *Metals Handbook*, volume 13, pages 104-122 and 583-609. Ninth edition, ASM International, Ohio, 1987.
10. Nisancioglu, K., Corrosion of aluminium alloys. *Proceedings of ICAA3*, volume **3**, pages 239-259. Trondheim, (1992). NTH and SINTEF.
11. Scamans, G. M.; Hunter, J. A.; N. J. H. Holroyd. Corrosion of Aluminum - a New Approach, *Proceedings of 8th International Light Metals Congress*, pages 699-705. Leoben-Wien, 1987.

Solar Wear Tattoo – Combining Functional Materials Taking Advantage of Different UPF of Textiles, to Create Healthier Tattooing Alternatives

Vanessa Mendes[1,a] and Mário Kong[2]

[1] *PhD Candidate, Faculty of Architecture, University of Lisbon – FA-UL, 1349-055 Polo Universitário – Alto da Ajuda, Lisboa, Portugal*
[2] *PhD Architect, Professor Faculty of Architecture - University of Lisbon, Researchers and CIAUD-permanent member of the Research Centre for Architecture, Urban Planning and Design FA-UL*

Abstract. Tattooing process can be a painful experience whether achieving a permanent or temporary body art on skin, and there can be horrible consequences. By taking advantage of UV protection factor variables of fashion clothing materials, it is possible to create this kind of body art in a much healthier way. The material performance is key, besides UPF materials that relies on the material behaviour when combined, layer sequence of these materials used for the desired function is equally as important, if not, even more important so. When properly combined in the correct order, making use of the right technology it may be possible to test different combinations. The advantages are, besides creating a temporary tattoo, that the user can benefit from moderate sun bathing advantages, that can provide the desired body art design on skin and the sunbathing health benefits. All of this can only be possible with a deep understanding of the UV protection factor of fashion materials, and how it can be tested and used in the manufacturing process for the desired function.

1 Body art – Tattooing process consequences

By body art it is understood a wide range of physical and aesthetic interventions on the body, the usual daily routine includes body art interventions such as hair styling [1] and waxing for both genders. [2] Tattoos are a kind of body art that through time revealed to have certain social [3] and health implications that come across as unwanted consequences of having a tattoo, whether the tattoo is permanent or temporary. The health consequences all come down to the choice of the type of tattoo. [4] [5]

1.2 Permanent tattooing process consequences

There are different types of tattoos and all have different techniques when it comes to marking the body. [6] It is possible to qualify them in two fundamental categories: Temporary or permanent.

Within the permanent category, there are loads of styles that assume different aesthetics due to the methods applied in the process of making the body modification. [7] When it comes to permanent

[a]Corresponding author: vanessa.g.mendes@hotmail.com

 DOI: 10.1051/matecconf/201710903001

tattoos or scarification, the main consequence is the shin damage made in order to make the desired tattoo design. [8] Other consequences include: unwanted scares due to allergies or other skin diseases. Both consequences can result from different skin sensitivities which may vary from individual to individual. [5]

Fig. 1 shows a skin disorder due to unwanted consequences of making a permanent tattoo. This infection can, and most likely will, result in unexpected and unwanted scarification on skin. [5]

Figure 1. Staphylococci infection with swelling of the entire arm and constitutional symptoms.

1.3 Temporary tattooing process consequences

The temporary category is essentially the *mehndi* practice. *Mehndi* is the art of painting the body with a *henna* solution. [9] The risk of this practice, is when this solution contains artificial colouring chemicals that can create a permanent scar due to contact ulcers that appear on the skin, in case of an allergy outburst. [10]

Fig. 2 shows an unwanted consequence of making a supposedly temporary tattoo that, most likely, became permanent due to unexpected scarification of the skin. [11] This allergic contact dermatitis that can leave residual depigmentation for weeks or even scarification, can be caused by the addition of synthetic dyes like para-phenylenediamine (PPD), para-toluylenediamine or related chemicals.
These chemicals are used to darken the *henna* colour and shorten application times, despite the consequences caused by the black *henna*. [4]

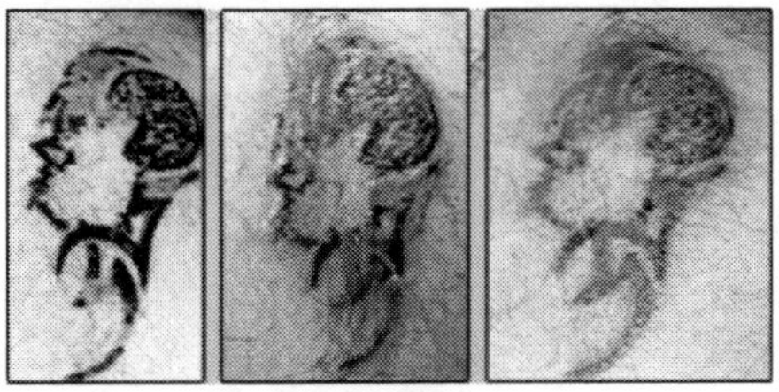

Figure 2. Example of skin contact ulcer due to skin allergy.

2 UV protection factor in clothes for tattooing purposes

Since every existing tattooing process can result in unwanted scars, a solution to this problem is to make use of an alleged 'textile disadvantage', to make tattooing through clothing possible. That will exclude the use of artificial pigments into making a tattoo, making use of the human body's natural pigment: melanin. [12]

Total UV protection textiles were made to provide total protection of the user's skin through clothes. These textiles were made because most textiles have different UV protection factors. These protection factors can change from textile to textile, depending essentially on composition [13] (including natural colour/ dying pigments), [14] [15] weave [16] and thickness of both yarn and finished textile. [17] Although UV protection factors in textiles can differ in less obvious ways by adding finishings to the textile, in order to enhance its UPF properties. [18] On the contrary, other textiles finishings such as mercerizing and canotizing, to make a fluorescent glow on the textile, can decrease its UPF properties. [19]

The more the body is covered up with high UPF textiles, the better the body will be protected from the sun. The burkini is a perfect example of 50+ UPF beachwear, [20] although high UPF factors are a main concern in everyday casual clothing. [21] These 2 examples imply that textiles with low UPF are a disadvantage.

Taking this information in to consideration, if different textiles and fashion materials with different UV protection factors are combined, it is possible to create contrasting two colour solar tattoos, making use of fashion materials and natural resources: Clothes and sun exposure. [22] Tanning beds are an alternative to sun exposure, as long as regulations are followed to guarantee the users health safety. The user in this case should also be informed that tanning beds are for tanning purposes only, [23] since just some tanning beds can provide vitamin D, not always being the case. [24]

2.1 Examples of varies of UV protection factors in textiles and how they can be used

2.1.1 The UV protection factors in textiles

Different textiles can have different UV protection factors according to their interaction with light. Taking in to consideration the material characteristics, they have different results when it comes to light interaction. It all comes down to reflection, absorption and transmission of the light. [25]

Table 1. Examples of textiles with different UV protection factors.

Sample details	UPF Rating	UV protection category
White Woven cotton	5+	Not relatable
White nylon knit	10+	Moderate
White polyester woven	15+	Moderate
White cotton knit	30+	Very high
White woven polyester crepe	50+	Maximum
Cream woven wool	50+	Maximum

The upper half of table 1 shows textiles that can be used to tan, since their protection factors range from low 5+ to moderate 15+.

The lower half of the same table shows textiles that can be used to block UVs, since their protection factors range from very high 30+ to maximum 50+. [26]

2.1.2 How they can be used

The essential rule to be applied, in order to achieve a solar tattoo through the use of fashion materials, is that the textiles and materials of the main area of the beachwear differ from the textiles and materials used on the tattooing area.

This is a product meant for the user. The fashion user has a need for multiple choices in order to satisfy its taste. [27] Changing materials variables is a delicate procedure, so dermatological tests may become required in order to ensure the users comfort, health and safety.

Meaning that comfort can affect the user when it comes to aesthetics but as well as the touch of the materials. [28] These materials need to be selected carefully and perhaps, may be in need of dermatological tests, in order to avoid complications that can create contact dermatitis. Allergic contact dermatitis can occur due to numerous chemicals that may be incorporated into the textiles and clothing during the processes form fibre formation, spinning, fabric construction to dyeing and finishing. [29]

2.2 Tattooing with fashion clothing materials – advantages and benefits

Sun exposure/ artificial UV exposure can result in several problems [30] but, it is proven that regular mild sun exposure can improve health and avoid physical issues related to vitamin D deficiency. [31]

2.2.1 Advantages/ benefits of creating tattoos with clothing materials

Sun exposure on a daily basis is, in fact, vital to a healthy body functioning. The clothing material area that allows the sun radiation to tan the users skin, will allow its body to receive the benefits that sun exposure can provide.

As well as the other area of the clothing that blocks the UV radiation on the user's skin, will benefit from the anti-ageing of the skin that is provided in the absence of sun exposure. [32]

It is up to the user to choose its beachwear model according to the desired tattoo design, and consequentially, the benefits he wants the most, considering his model of choice, that can block most of the UV rays from the skin, making a tattoo area based on melanin stimulation, or, allowing most of the UV rays on the skin, making a tattoo area based on melanin stimuli restriction.

3 Healthier body art tattoo – How it works

The textiles characteristics mentioned can be combined to create an efficient and healthy tattoo, with all the benefits of sun exposure. The key is to put these materials in the right places meaning that, if the desire is to make a positive tattoo, materials that are UV receptive are used in the tattoo area and to make a negative tattoo, materials that block UV are used in the tattoo area.

3.1 Previous tests results

Previous tests were made with textiles, providing the information that different materials with different material behaviours can, create contrasting results on skin, using various fashion materials with different UPF. The materials used were 2 knitted textiles made of polyamide, with different structures but, similar in colour and thickness. Additional fashion adornments were added as well.

Material overlap on its own is enough to create a possible mark on skin, due to the amplified thickness. So, in order to remove the possibility of unwanted marks on skin, seam overlay can be hidden in the material overlap area of UV blockage. (Figure 3) [22]

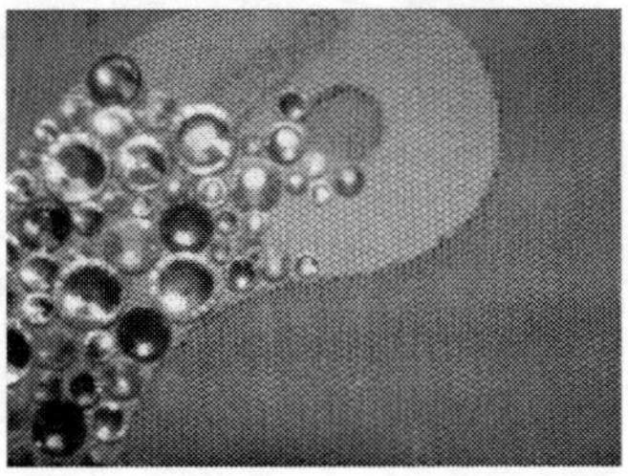

Figure 3. Materials and finishings overlay with the seam in between – Authors photograph.

Despite the overlap used, it is not essential for the UV blockage since this blockage works on its own without the seam overlay, it is only an aesthetic improvement to avoid unwanted sun marks due to bad pattern making, that can expose a seam textile overlay to UV, making it possible to appear an unwanted tan mark on the user's skin as a consequence (Figure 4) [22]

Figure 4. Materials and finishings overlay without the seam in between. – Authors photograph

In a popular sense, the user tends to call a bad tan line to an anaesthetic lace silhouette on skin, as a result of bad beachwear design. [33] In this sense, an exposed seaming overlay in low SPF textiles could be called as bad beachwear design, and should be avoided by hiding it in the materials and finishings overlap.

Alternative to this manufacturing process, it's to use fabric/ outfit made in a circular loom. A process that is usually used to make socks without seams. [34] This way, eliminating the sewing area, there is no need to hide or incorporate it in the UV blockage area, in order to prevent the so called bad tan line. This manufacturing process should be included in future prototypes. In this case, extra care with the knitwear should be taken, due to the knitwear behaviour.

Knitted structured fabrics stretch capacity, is part of the physical behaviour of its knit loops. [35] Material overlap is key to improve UPF in the tattooing area, increasing thickness and raising UPF. The strategical overlap of these materials creates a stronger UV barrier. Plus, the overlap of the materials used, restrains the natural stretching of the knitted fabric structures. (Figure 5)

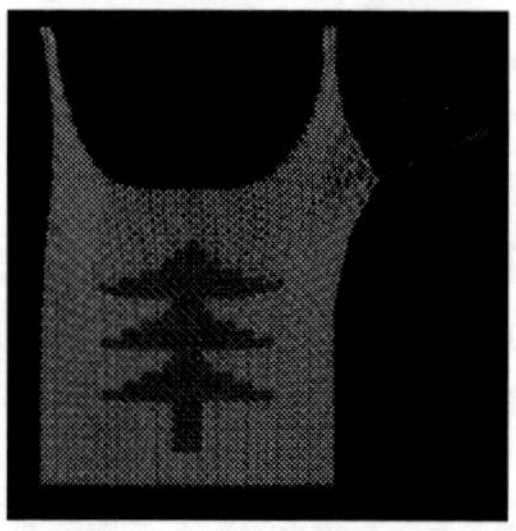

Figure 5. Physical behaviour of knitwear. [35]

Restraining the knit loops natural behaviour, prevents the tattoo area distortion. [22] The restrain area can go in the tattoo, to make a negative tattoo (Figure 6) or, in the area that surrounds the positive tattoo, to prevent the distortion of the positive tattoo area. Further research is needed in order to draw conclusions for the positive solar tattoo.

Figure 6. Knitted loops restrain. – Authors photograph.

3.2 Other materials and techniques – Same logic

The same logic can be applied with different materials, that can lead to similar results. The goal is to always create a contrasting skin colour through melanin stimuli, making use of fashion materials with different UPF. The manufacture can also suffer changes. These changes can be applied specially to the sequence in which materials and finishings are overlapping and firming up together, from the inner layers that come in contact with the user's skin, to the outer layers that are exposed to external elements.

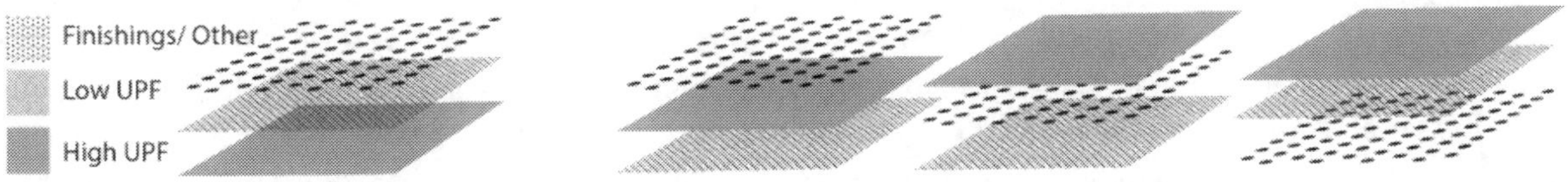

Figure 7. Materials and finishings overlay sequences diagrams with labels.

In Fig. 7, the bottom section represents the materials or finishings that come into contact with the user's skin. The upper section of the same figure (Figure 7) represents the materials or finishings of the outer layer that stay on the surface, on the show, and can be directly observed while the user is wearing a finished outfit/ accessory made according to the respective material overlay sequence.

On the far left, materials and finishings overlay sequence used in the previous test.

On the right, examples of materials and finishings overlay sequences to be tested in the present ongoing research. Materials and finishings overlay sequences, other than the ones that are exemplified above, may be tested as well.

The outer layers need extra attention when it comes to materials resistance to natural degradable factors, since UV exposure, of both artificial light of tanning beds and natural sun light, can damage the textiles properties. [13]

Other textile damaging agents can be found in places that user may go to tan his skin. Swimming pool water contains chlorine which affects the textiles durability with repeated exposure. [36]

Whatever the materials and finishings overlay sequences chosen, it is expected a solar tattoo to appear on skin, due to contrast between melanin stimuli and UV blockage.

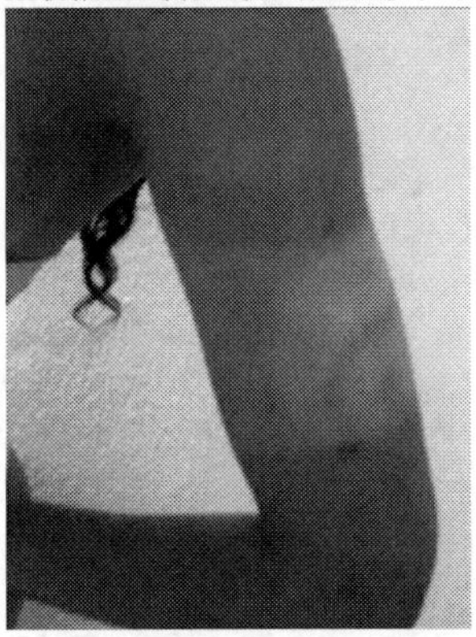

Figure 8. Negative solar tattoo on skin, making use of the previous test prototype – Authors photograph [22]

The temporary tattoo obtained through sun exposure, (Figure 8) is an example of the desired results of contrast on skin.

Other prototypes made with other materials, finishings and overlay sequences, will be tested with the goal of creating a contrasting drawing in the body. This contrast can be achieved like this, by creating a tattoo area in negative with melanin stimuli restriction, or the other way around, allowing to tan in the drawing area, making a positive tattoo, blocking the UV rays in the rest of the prototype.

No quantitative data was registered in this prototype's first test. Tests may need to be repeated in order to retrieve precise information about UPF of these materials, in order to be able to compare the results with the future prototypes to be made and tested as well.

4 Conclusions

Spectrometers (and other accessories) are essential tools to calculate the spectral region, measuring the UV transmission in fabrics. [26]

Further research is needed to fully understand the limits of the range of materials that can be combined, and their orders of overlap sequences respectively. Samples should be made and tested by measuring the UV transmission with the necessary equipment, to obtain quantitative results with precision.

Test results can be compared, so that full understand the potential of each sample can be achieved. Only then it is possible to decide which sample is the most effective for the desired result. Creating a high contrast on the skin, due to samples that combine textiles with different UPF.

The main reason to make different multiple tests, besides obtaining effectiveness results, testing the limits of material combinations that allow for this solar tattoo through textiles, opens up to new possibilities, which is important in order to satisfy the users need of choice.

This practical experiment can be achieved through the application of a quantitative experimental methodology, with a qualitative research basis on the users taste, [37] to expand the technological possibilities of the present ongoing research.

References

1. T. Nicolas, *Body art*, 126 (Thames & Hudson, London, 2014)
2. A. Smelik, Cri. Stu. Fash. Bea., *A close shave: The taboo on female body hair*, **2,** 240 (2015)
3. M. Miflin, *Bodys of subversion*, 134, 136 (powerhouse books, Brooklyn, New York, 2013)
4. A. Hazra, Ind. Jou. Psic. *XXXIV Annual conference of the Indian pharmacological society, Adverse reactions to henna*, **34**, 436 (2002)
5. J. Serup, J. T. Linnet, O. Olsen, N. Harrit, B. Møhl, H. Westh, *The council on health and disease prevention, Tattoos – Health, risks and culture* 70, 87 (2015)
6. E. Schildkrout, Ant. Win., *Body art as visual language*, **2**, 3, 4 (2001)
7. S. M. Abbasi, Ind. J. Res. Par., *Body adornment and modification: Is it art or is it torture*, **2**, 1, 2 (2013)
8. R. Lemay, *The tattoo artist's black book*, 36 (2008)
9. C. Fabius, *Mehndi*: *The art of henna painting* (Harmony, 1998)
10. S. Benomar, L. Ismaili, M. Rmili, L. Benzekri, B.; Hassam, *Revue Française d'Allergologie, Tatouage au henné noir: au-delá de l'eczéma de contact*, **49,** 376-3778 (Elsevier B. V., 2009)
11. C. Cartwright-Jones, *Henna is not black* (2003)
12. F. Hu, J. Soc. Cosm. Chem., *Melanocytes and melanin pigmentation*, **19**, 565-580 (1968)
13. D. Saravan, AUTEX Res. J., *UV protection textile materials*, **7**, 56 (2007)
14. S. Venkatakrishnan, P. K. Mandhyan, R. Narkar, M. K. Kumar, Ind. J. App. Res., *Studies on the physical properties of naturally coloured cotton & development of eco-friendly UV protected fabric*, **5**, 643 (2015)
15. A. A. Salman, Z. M. Abdel-Megied, N. El-Mohsen, Seddik K. M., M. El-Aziz, Int. J. Sci. Res., *Textiles & apparel as UV protectives in medical applications*, **4**, 2302 (2015)
16. A. M. Grancarić, Ž. Penava, A. Tarbuk, *VI Symposium contemporary technologies and economic development, UV protection of cotton – The influence of weaving structure*, 230 (2005)
17. P. D. Dubrovski, *Woven fabrics and ultraviolet protection*, 284 (Intech, Croatia, 2010)
18. A. Ghani, W. Qorshi, A. Romyia, R. El-Sayd, Int. Des. J., *Treatment of children's garments natural fabrics for protection against ultraviolet radiation*, **4**, 104 (2010)
19. A. M. Grancarić, T. Pušić, A. Tarbuk, I Jančijev, Int. Col. Ass., I*nterim Meeting of the International Color Association*, *The fluorescence of sunprotected white cotton fabrics*, 126 (2004)
20. K. Dabrowska, *The burkini: A muslim sportswomens' dream come true*, **29** (Islamic Tourism, 2007)

21. P. Hulme, *Fabrics China® Hangtag Program Gains Momentum*, **5** (Textile connection, Singapore, 2014)
22. V. Mendes, *Solar Wear Tattoo* (Faculty of Architecture, Lisbon, 2015)
23. H. W. Lim et al, J. Am. Acad. Dermatol., *Sunlight, tanning booths and vitamin D*, **52**, 871-872 (2009)
24. S. Balk, D. Fisher, A. Geller, M. Weinstock, *Evaluation of indoor tanning health claims* (State Office of the Attorney General, New York, 2015)
25. H. Yang, S. Zhu, N. Pan, J. App. Pol. Sci., *Studying the mechanisms of titanium dioxide as ultraviolet-blocking additive for films and fabrics by an improved scheme*, **92**, 3202 (2003)
26. Z. Bilimis, *Measuring the UV protection factor (UPF) of fabrics and clothing* (Agilent technologies, Mulgrave, Australia, 2011)
27. G. Moore, *Promoção de moda*, 20 (Editorial Gustavo Gili, Barcelona,2013)
28. S. Bragança, L. Fontes, P. Azeres, E. R. Edelman, M. Carvalho, *6th international conference on applied human factors and ergonomics*, *The impact of work clothing design on workers' comfort*, **3**, 5890 (Elsevier B. V., 2015)
29. W. Zhong, *Cutaneous and ocular toxicology*, *Textiles and human skin, microclimate, cutaneous reactions: An overview*, **25**, 25 (Taylor & Francis, 2006)
30. J. D'Orazio, S. Jarrett, A. Amaro-Ortiz, T. Scott, Int. J. Mol. Sci., *UV radiation and the skin*, **14**, 12226, 12227 (2013)
31. T. C. Chen, Z. Lu, M. F. Holic, *Photobiology of vitamin D*, 36, 37 (Humana press, 2010)
32. M. Ichihashi, H. Ando, M. Yoshida, Y. Niki, M. Matsui, *Anti-Ageing Medicine*, *Photoageing of the skin*, **6**, 53, 54 (Japanese Society of Anti-Ageing Medicine, 2009)
33. T. E. Morissei, *The strappy swimsuit trend creates the world's worst tan lines* (2016) [http://jezebel.com/the-strappy-swimsuit-trend-creates-the-worlds-worst-tan-1585463004]
34. M. Araújo, E. M. M. Castro, *Manual de engenheiria textil* (Calouste Gulbenkian Foundation, Lisbon, 1986).
35. F. Durupinar, U. Güdükbay, *Computer & Graphics*, *Procedural visualization of knitwear and woven cloth*, **31,** 781 (Elsevier B. V., 2007)
36. K. M. Borchard, *California state science fair, Polyester vs. Nylon: Which material is best for swimsuits?* (2006)
37. L. Pardal, E. S. Lopes, *Métodos e técnicas de investigação social*, 20, 21 (Areal Editores, Porto, 2011)

Fabrication of Polycrystalline Transparent Co^{+2}: $MgAl_2O_4$ by a Combination of Spark Plasma Sintering (SPS) and Hot Isostatic Pressing (HIP) Processes

Maxim Sokol[a], Sergey Kalabukhov and Nahum Frage

Department of Materials Engineering, Ben-Gurion University of the Negev, P.O.B 653, Beer-Sheva 84105, Israel

Abstract. Transparent Co^{2+} doped $MgAl_2O_4$ spinel was fabricated by SPS consolidation followed by and HIP treatment. It was established that HIP treatment significantly improved transparency of the ceramic in a wide range of wavelengths, especially, in a range, which is relevant for Q-switching. Nonlinear absorption was demonstrated and the ground and excited state absorption cross sections were estimated. The positive effect of the HIP treatment on the optical properties is related to an elimination of extremely fine porosity and to the location of Co ions at Mg^{2+}sites in the spinel ionic structure. The experimental results indicate that the fabricated specimens can be used as a passive laser Q-switching material.

1 Introduction

Most types of passive laser Q-switching materials developed in the last two decades are based on transition metals cations like Co^{2+}, V^{3+}, or Cr^{4+} as an active dopant [1–4]. For infrared lasers in the 1.3–1.7   m range, especially for lasers exploiting the $^3I_{13/2} \rightarrow\ ^4I_{15/2}$ emission, passive Q-switches based on Co^{2+} doped $MgAl_2O_4$ spinel are an efficient solution [5–13]. Methods of the fabrication and optical properties of Co^{2+} doped $MgAl_2O_4$ single crystals [5, 7, 9] or glass-ceramics [14-15] have been widely studied and discussed in literature. Conventional methods are expensive and time consuming. Recently, SPS [16] and HIP [17] have been successfully applied to fabricate polycrystalline transparent Co doped $MgAl_2O_4$. The SPS process is very fast and may be carried out at relatively low sintering temperatures; however, it doesn't always provide a desired transmittance of ceramics. Prolonged treatments in the SPS apparatus lead to a massive interaction between graphite tooling and $MgAl_2O_4$. It was suggested that a combination of a short SPS process and additional HIP treatment may make available the fabrication of the transparent ceramic with improved optical properties.

2 Experimental Procedures

2.1 Powder synthesis and sintering

Co-doped spinel powder containing 0.09 at. % Co was used as starting material. The powder was synthesized at the Israeli Ceramic and Silicate Institute, Haifa, Israel by wet-chemistry, based on

[a] Corresponding author: sokolmax@post.bgu.ac.il

hydroxides co-precipitation as described in details by A. Goldstein et al. [17]. The Co-doped spinel powder was premixed with 0.7 wt.% LiF (99.98%, Alfa-Aesar, Heysham, Lancashire, UK). The mixing procedure was described in details in our previous contributions [18, 19]

The fabrication process was done in two steps. At the first step, consolidation experiments were performed using an SPS apparatus (FCT Systems, Rauensin, Germany). SPS parameters were as follows: the sintering temperature was 1550ºC, the heat rate was 20 ºC/min, the dwell time was one hour and the applied pressure was 60 MPa. After the SPS treatment the specimens were grinded and polished to an optical quality surface. Finally, the specimens were HIPed for 4 h at 1550ºC under 200 MPa Ar gas pressure.

2.2 Microstructure, mechanical and optical properties

Micro-structure of the SPS/HIP processed specimens was analysed by an inverted metallography light microscope (Axio, Zeiss, Oberkochen, Germany). Grain size distribution was determined from the micrographs using Thixomet image analysis software [20-22]. Optical in-line transmission measurements were performed using Bruker Vertex V70 FTIR in wavelength range between 0.8 and 7 μm. Bending strength was determined by a 3-point test using a LRXPlus tensile tester (Lloyd Instruments, Fareham, U.K.). Vickers hardness measurements were conducted applying 2000 g load by a Buehler Micromet 2010 apparatus. The elastic modulus was determined by the "pulse echo" method.

2.3 Nonlinear saturation properties

The transmission saturation measurements were performed by Nd: YAG laser with relatively high peak power as a pump laser for an optical parametric oscillator (OPO). A tuneable Potassium titanyl phosphate (KTP) crystal parametrically converts the laser wavelength from the 1065 nm pump to 1417 nm (tuneable OPO) signal. The signal was separated from the pump and idler beams by two filters and focused by a lens, allowing determining the beam waist by using the knife-edge technique. Finally, the transmitted radiation through the Co^{2+}: $MgAl_2O_4$ specimen was measured by a PbSe photo diode power meter.

3 Result and Discussion

The obtained specimens after SPS process were fully dense and translucent with dark tint between blue to black colours. An additional HIP treatment significantly improved the transparency. Figure 1 shows a photograph of the optically polished 2 mm thick Co^{2+} doped $MgAl_2O_4$ spinel sample after HIP treatment.

Figure 1. Co^{2+}: $MgAl_2O_4$ sample fabricated by SPS and HIP processes.

3.1 Microstructure and mechanical properties

Microstructure of the fully dense spacemen is presented in Figure 2. The average grain size was determined to be 8.2 microns.

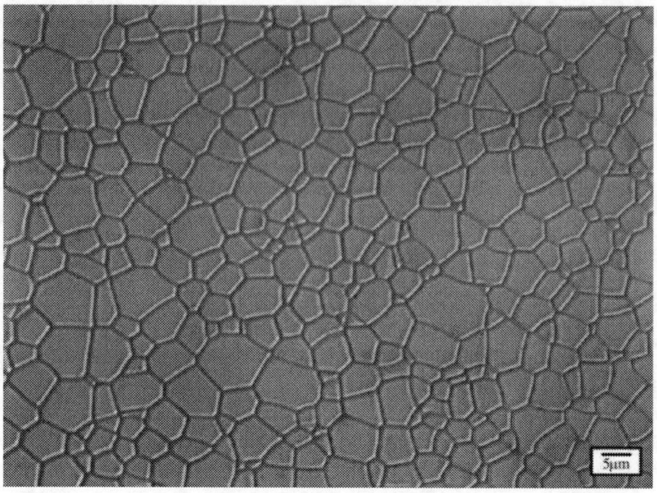

Figure 2. Microstructure of Co^{2+}: $MgAl_2O_4$ sample fabricated by SPS and HIP processes.

The values of hardness, bending strength and Young modulus for the cobalt doped ceramic are presented in Table 1. The obtained mechanical properties are in a good agreement with those previously reported for un-doped spinel of the same grain size [23, 24].

Table 1. Mechanical properties of polycrystalline Co^{2+}:$MgAl_2O_4$.

Vickers hardness, HV	Bending strength, MPa	Young modulus, MPa
1330±30	220±11	285±5

3.2 Optical properties

The optical transmittance from visible to the mid-IR and short-IR wavelength range is presented in Figure 3. In the visible range (at wavelength of 450 nm) a transmittance is about 70%, while at the mid-IR to short-IR spectra, which is the relevant rang for Q-switching the transmittance is higher than 80% and is closed to the theoretical value.

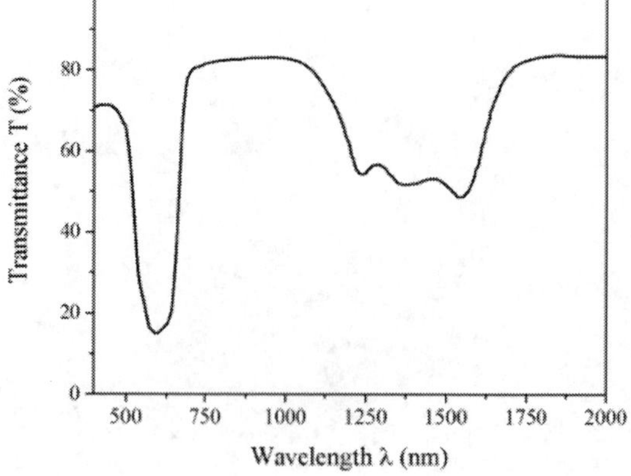

Figure 3. Transmission spectrum of a 2 mm thick Co^{2+}:$MgAl_2O_4$ sintered by SPS followed by HIP treatment.

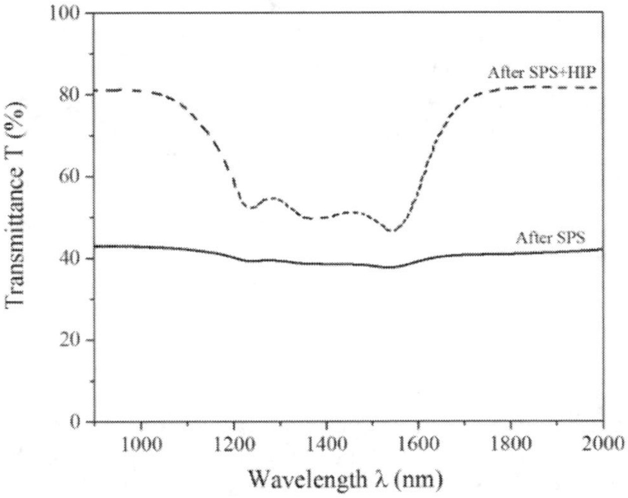

Figure 4. Transmission spectrum of a 2 mm thick Co^{2+}:$MgAl_2O_4$ of the SPS-processed and specimens after additional HIP treatment.

The effect of HIP treatment on the transparency is presented in Figure 4. For the SPS processed specimen the transmittance is relatively low for the entire range of wavelengths, for the HIPed specimen the transparency was significantly higher and a strong absorption related to the $^4A_2(^4F)\rightarrow{}^4T_1(^4F)$ transition [17] of the Co^{2+} ions was observed. The positive effect of HIP treatment on the spectra may be related to two factors: additional pore elimination by the high isostatic pressure and diffusion of Co ions to T_d symmetry sites of Mg^{2+} in the spinel structure [25] as a result of the additional dwell time at high temperatures.

3.3 Nonlinear saturation properties

The dependence of incident fluence on the transmission of Co^{2+}:$MgAl_2O_4$ at a wavelength of 1417nm is presented in Figure 5.

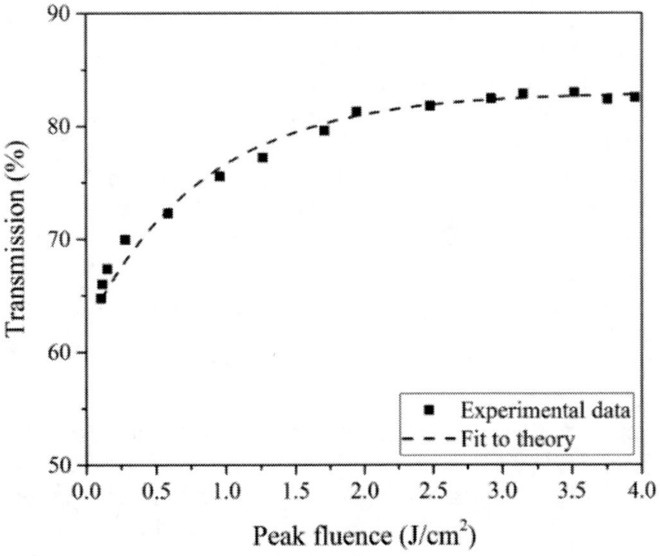

Figure 5. Transmission as a function of the incident fluence at 1417 nm for Co^{2+}:$MgAl_2O_4$ sintered by SPS followed by HIP treatment.

To measure the ground and excited state absorption cross sections through saturable absorption experiments, the energy fluence was increased until the excited state level was full with electrons. At this state the transmission curve becomes asymptotic. The results were fit to a slow saturable absorber model [11], when the pulse duration is much shorter than the excited state decay. The ground-state and the excited-state absorption (σ_{gsa} and σ_{esa}) cross-sections and the Co^{2+} ion density N were calculated from the fit. The estimated values are found to be $\sigma_{gsa} = 3.57 \cdot 10^{-19}\, cm^2$, $\sigma_{esa} = 2.49 \cdot 10^{-20}\, cm^2$ and $N = 1.05 \cdot 10^{19}\, cm^{-3}$, respectively. These values are in good agreement with previously reported results for single crystal [7], SPS processed [16] and HIP processed [17] Co-doped spinel.

4 Conclusions

The combination of SPS and HIP processes was applied for the fabrication of the Co^{2+} doped $MgAl_2O_4$ ceramic. The additional HIP treatment allowed significantly improving transmittance of the ceramic in a wide range of wavelengths. For the wavelengths range, which is the relevant, rang for Q-switching the transmittance is higher than 80% and is closed to the theoretical value. Nonlinear absorption was demonstrated. The ground and excited state absorption cross sections were estimated according to slow saturable absorber model. It was suggested that the positive effect of HIP treatment on the spectra is related to an elimination of extremely fine porosity and to the location of Co ions at sites of Mg^{2+} in the spinel ionic structure.

References

1. R. Feldman, Y. Shimony, and Z. Burshtein, Opt. Mater. 24 333–344 (2003).
2. R. Feldman, Y. Shimony, and Z. Burshtein, Opt. Mater. 24 393–399 (2003).
3. P. Li, X. H. Chen, H. N. Zhang, and Q. P. Wang, Laser Phys. 21 1708–1711 (2011).
4. M. B. Camargo, R. D. Stultz, M. Birnbaum, and M. Kokta, Opt. Lett. 20 339–341 (1995).
5. D. P. Jiang, Y. Q. Zou, L. B. Su, H. L. Tang, F. Wu, L. H. Zheng, H. J. Li, and J. Xu, Laser Phys. Lett. 8 343–348 (2011).
6. Y. V. Volk, A. M. Malyarevich, K. V. Yumashev, O. S. Dymshits, A. V. Shashkin, and A. A. Zhilin, J. Appl. Spect. 74 140–146 (2007).
7. Y. V. Volk, A. M. Malyarevich, K. V. Yumashev, V. N. Matrosov, T. A. Matrosova, and M. I. Kupchenko, Appl. Phys. B 88 443–447 (2007).
8. K. V. Yumashev, Appl. Opt. 38 6343–6346 (1999).
9. K. V. Yumashev, I. A. Denisov, N. N. Posnov, N. V. Kuleshov, and R. Moncorgé, J. Alloy. Compd. 341 366–370 (2002).
10. R. M. Boiko, A. G. Okhrimchuk, and A. V. Shestakov, Advanced Solid State Lasers, OSA Trends in Optics and Photonics Series, paper LS9 (1998).
11. Y. V. Volk, I. A. Denisov, A. M. Malyarevich, K. V. Yumashev, O. S. Dymshits, A. V. Shashkin, A. A. Zhilin, U. Kang, and K. H. Lee, Appl. Optics 43 682–687 (2004).
12. A. Denisov, Yu. V. Volk, A. M. Malyarevich, K. V. Yumashev, O. S. Dymshits, A. A. Zhilin, Uk Kang and Kyeong-Hee Lee, J. Appl. Phys. 93 3827–3831 (2003).
13. O. Dymshits, A. Shashkin, A. A. Zhilin, Y. Volk, A. Malyarevich, and K. Yumashev, Adv. Mater. Res. 39–40 219–224 (2008).
14. A. M. Malyarevich, I. A. Denisov, K. Yumashev, O. S. Dymshits, A. A. Zhilin, and U. Kang, Appl. Opt., 40, 4322–5 (2001).
15. S. Y. Feng, C. L. Yu, L. Chen, S. G. Li, W. Chen, and L. L. Hu, Laser Phys., 20 [8] 1687–91 (2010).
16. A. Wajler, A. Kozłowska, M. Nakielska, K. Leśniewska-Matys, A. Sidorowicz, D. Podniesiński and P. Putyra, J. Am. Ceram.Soc., 97 [6] 1692–5 (2014).

17. A. Goldstein, P. Loiko, Z. Burshtein, N. Skoptsov, I. Glazunov, E. Galun, N. Kuleshov, K. Yumashev, J. Am. Ceram. Soc., 99 [4] 1324-1331 (2016).
18. N. Frage, S. Kalabukhov, N. Sverdlov, V. Ezersky, M. P. Dariel, J. Eur. Ceram. Soc., 30 [16] 3331–7 (2010).
19. N. Frage, S. Kalabukhov, N. Sverdlov, V. Kasiyan, A. Rothman, M.P. Dariel, Ceram. Int., 38 [7] 5513–9 (2012).
20. A. Kazakov, N. Luong, E. Kazakova, E. Zorina, Microstruct. Sci. 27 133–142 (2000).
21. A. Kazakov, N.H. Luong, Mater. Charact. 46 155–161 (2001).
22. A. Kazakov, D. Kiselev, Microstruct. Anal. 1–8 (2016)
23. A. Krell and A. Bales, App. Ceram. Tech., 8 1108-1114 (2011).
24. A. Rothman, S. Kalabukhov, N. Sverdlov, M. P. Dariel and N. Frage, Int. J. App. Ceram. Tech., 11 146-153 (2014).
25. T. Bates, *Modern Aspects of the Vitreous State*, (Butterworth, London, 1962)

In Vitro Bioactivity and Setting Times of White Portland Cement Combined with Different Radio Pacifying Agents

Nichola Jayne Coleman[1,a], Rajitha Hanarasinghe[2], Zeynep Aslı Güçlü[3] and Samantha Ellen Booth[1]

[1]*University of Greenwich, Faculty of Engineering and Science, Chatham Maritime, Kent, ME4 4TB, UK*
[2]*Faculty of Medicine, University of Peradeniya, Galaha Road 20400, Peradeniya, Sri Lanka*
[3]*Department of Paediatric Dentistry, Erciyes University, 38039 Kayseri, Turkey*

Abstract. Commercial formulations based on 80:20 mixtures of Portland cement and bismuth oxide (a radiopacifying agent) are used in dentistry as root-filling materials. This study compares the impact of two alternative radiopacifiers, barium sulphate and zirconium oxide, with that of bismuth oxide, on the setting times and bioactivity of white Portland cement. The findings indicate that bismuth oxide prolongs both the initial and final setting times of the cement, and that barium sulphate and zirconium oxide have no effect on this parameter. Hydroxyapatite (HA) formed on the surfaces of all test samples within 7 days of exposure to simulated body fluid, indicating that they possess the potential to stimulate new hard tissue formation. Fourier transform infrared spectroscopy, the traditional technique for the identification of HA, was not appropriate for the analysis of these cement systems owing to the overlap of signals from each of the radiopacifiers with the characteristic P-O bending modes of HA in the 570 – 610 cm^{-1} region. In this respect, the P-O band at 965 cm^{-1} of HA in the Raman spectrum was found to be a suitable means of detection since it is discrete with respect to all signals arising from the radiopacifying agents and cement phases.

1 Introduction

Mineral trioxide aggregate (MTA) is a commercial dental cement comprising a mixture of 80 wt% Portland cement and 20 wt% bismuth oxide (Bi_2O_3), a radiopacifying agent, which has been used in endodontic restorations for the past two decades [1]. Bismuth oxide does not directly engage in the hydration reactions of Portland cements; however, its physical presence in the cement system is reported to retard the hydration processes, to increase the porosity of the cement matrix and to decrease mechanical stability [2, 3]. Bismuth oxide is also reported to discolour dental tissues [4].

Barium sulphate ($BaSO_4$) and zirconium oxide (ZrO_2) are among a range of alternative radiopacifying agents with potentially superior chemical, mechanical and biological properties that are currently under investigation for use in hydraulic calcium silicate cements for dental restorations [4-7]. Both barium sulphate and zirconium oxide, by virtue of their low solubilities, are currently used as radiographic contrast agents in a number of biomaterials [5, 8].

In this study, experimental cements were prepared with 80 wt% white Portland cement (WPC) and 20 wt% of either barium sulphate or zirconium oxide. The in vitro bioactivities of these experimental cements were tested by exposure to simulated body fluid [9], and compared with those of unblended

[a] Corresponding author: n.coleman@gre.ac.uk

WPC and WPC blended with 20 wt% bismuth oxide (i.e. the proprietary radiopacifier). The impact of the various radiopacifiers on initial and final setting times were also determined using a Vicat apparatus.

2 Materials and methods

2.1 Materials and sample preparation

The WPC used in this study was supplied by Lafarge [8]. Anhydrous barium sulphate, bismuth oxide, zirconium oxide and hydroxyapatite (HA), ex. Sigma Aldrich, UK, were used without further purification. Cement paste samples were prepared by mixing 10 g of WPC with 3.5 g of distilled water. The specimens were sealed into polypropylene tubes and cured at 37 °C (i.e. body temperature) for 24 hours. Samples blended with barium sulphate, bismuth oxide or zirconium oxide (viz. WPC-Ba, WPC-Bi and WPC-Zr, respectively) were prepared similarly with 20 wt% replacement of WPC for radiopacifier (at the same water:cement ratio of 0.35 by mass). Each specimen was prepared in duplicate.

2.2 Initial and final setting times

Initial and final setting times of the cement pastes were determined by the method described in ASTM C191-08 using a manual Vicat apparatus [8]. Setting data for each cement were collected in triplicate and were subjected to a two-tailed *t*-test at $P = 0.05$.

2.3 In vitro bioactivity

Simulated body fluid (SBF) was prepared by the method described in reference 9. 0.15 g of cement sample were immersed in 150 cm^3 of SBF in sealed polypropylene containers at 37 °C for 3, 6, 24, 48, 72 and 168 hours. Each analysis was carried out in triplicate. The supernatant solution concentrations of Si, Al, Ca, P, Ba, Bi and Zr species were determined by inductively coupled plasma analysis (ICP) using a TJA Iris simultaneous ICP-OES. The recovered cement samples were rinsed with deionised water, dried in air at 37 °C for 24 h and analysed by Fourier transform infrared spectroscopy (FTIR) using a Perkin Elmer Paragon spectrometer and Raman spectroscopy using a LabRam I Raman instrument fitted with an Olympus BX 40 microscope at 633 nm wavelength and a spot size of 5 µm. FTIR and Raman spectra were also collected for synthetic hydroxyapatite, each radiopacifying agent, and the cement samples (WPC, WPC-Ba, WPC-Bi and WPC-Zr) after curing for 7 days at 37 °C.

3 Results and discussion

3.1 Initial and final setting times

The initial and final setting times for samples WPC, WPC-Ba, WPC-Bi and WPC-Zr were obtained using a manual Vicat apparatus in accordance with ASTM C191-08, and are listed in Table 1.

Table 1. Initial and final setting times for WPC, WPC-Ba, WPC-Bi and WPC-Zr.

Cement	WPC	WPC-Ba	WPC-Bi	WPC-Zr
Initial set (min)	150 ± 15	160 ± 10	255 ± 15	165 ± 5
Final set (min)	200 ± 17	197 ± 6	345 ± 28	198 ± 8

These data demonstrate that the presence of 20 wt% bismith oxide significantly increases both the initial and final setting times of WPC by 105 min and 145 min, respectively; whereas, barium sulphate and zirconium oxide have no impact on either setting time. These findings confirm those of other studies that also report an increase in setting time associated with bismuth oxide [2, 3]. A decrease in setting times has previously been reported for the same white Portland cement containing 20 wt% zirconium oxide at a water: cement ratio of 0.375 [8]. The apparent discrepancy between these findings may be attributed to the lower water: cement ratio used in this study.

The ability of a radiopacifying agent to increase setting time is a distinct disadvantage, as a common problem regarding MTA is that its setting time (which is reported to vary between 50 min and 2 h 45 min) is already inconveniently long from a clinical perspective [10]. In this respect, barium sulphate and zirconium oxide are both superior to bismuth oxide, the proprietary radiopacifying agent.

3.2 In vitro bioactivity

The formation of a surface layer of substituted HA, ($Ca_{10}(PO_4)_6(OH)_2$), on a material placed in SBF solution provides an indication of its bioactivity (i.e. the ability of the material to bond with living bone tissue and to subsequently stimulate tissue regeneration) [9].

The FTIR spectra of HA, barium sulphate, bismuth oxide and zirconium oxide are shown in Fig. 1a. These FTIR spectra demonstrate that various vibrational modes of each of the radiopacifiers overlap with the characteristic P-O bending mode of crystalline HA at 570 - 610 cm^{-1}, which is commonly used to identify deposits of HA on the surface of materials after exposure to SBF [11].

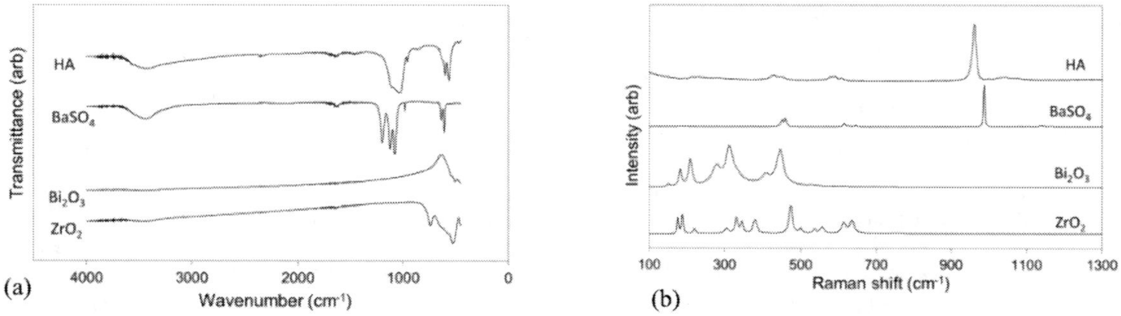

Figure 1. (a) FTIR spectra; and (b) Raman spectra of hydroxyapatite, barium sulphate, bismuth oxide and zirconium oxide.

The Raman spectra of HA, barium sulphate, bismuth oxide and zirconium oxide are shown in Fig. 1b. The Raman spectrum of each of these mineral phases closely resembles those in the literature [12-15]. These spectra demonstrate that the phosphate stretching mode of HA at 965 cm^{-1} is discrete with respect to all Raman-active bands arising from the radiopacifying agents.

The FTIR and Raman spectra of the experimental cements cured for 7 days are shown in Fig. 2, and those of the cements that were immersed in SBF for 7 days are presented in Fig. 3.

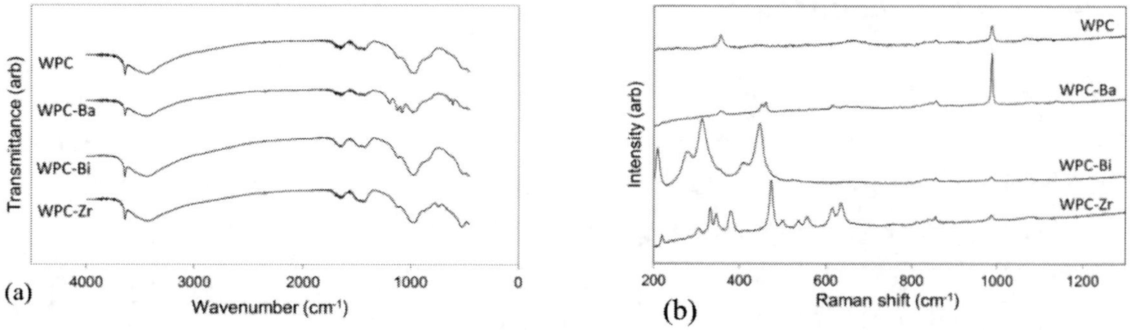

Figure 2. (a) FTIR spectra; and (b) Raman spectra of samples WPC, WPC-Ba, WPC-Bi and WPC-Zr, after curing for 7 days at 37 °C.

In the FTIR spectra of the cured cements (Fig. 2a), the band at ~970 cm^{-1} is assigned to Si–O stretching vibrations of the C-S-H gel cement hydration product, with contributions at 870 and 1120 cm^{-1} from carbonate and sulphate groups, respectively [11]. Carbonate species also give rise to the broad unresolved doublet centred at 1470 cm^{-1} and the signals at 1660 and 3470 cm^{-1} are attributed to O–H vibrations of water and hydroxyl groups of various hydrated cement phases. The sharp signal at 3670 cm^{-1} is assigned to the stretching mode of O–H in the crystalline calcium hydroxide hydration product. Bands arising from the various vibrational modes of the radiopacifying agents are noted to contribute to the FTIR spectra of the blended cement samples and are particularly apparent in the 550 - 630 cm^{-1} region.

Raman signals of most hydrated cement phases are inherently weak, and so the spectra of hydrated Portland cements tend to be dominated by bands arising from sulphate bending and stretching modes of unreacted gypsum and ettringite [16]. In the Raman spectra of the cured cements (Fig. 2b), the bands at 986 and 853 cm^{-1} arise from sulphate stretching modes and the band at 354 cm^{-1} is assigned to sulphate bending [16]. The signals arising from the radiopacifiers in the Raman spectra of the blended cements are considerably stronger than those of the cement phases (Fig. 2b).

The FTIR spectra of the cement samples following immersion in SBF are shown in Fig. 3a. In the spectrum of WPC, the doublet at 570 and 607 cm^{-1} is indicative of the formation of crystalline HA which confirms the bioactivity of this material. The dissolution of calcium hydroxide is also demonstrated by the absence of the sharp O–H signal at 3670 cm^{-1} following immersion in SBF. In the FTIR spectra of the blended cements, it is difficult to determine whether HA deposition has occurred, as bands from the radiopacifying agents obscure the region in which the characteristic P-O signal of HA appears (Fig. 3a). Conversely, the characteristic Raman-active phosphate stretching band of HA at 965 cm^{-1} is clearly and unambiguously observed in the Raman spectra of all samples following immersion in SBF (Fig. 3b).

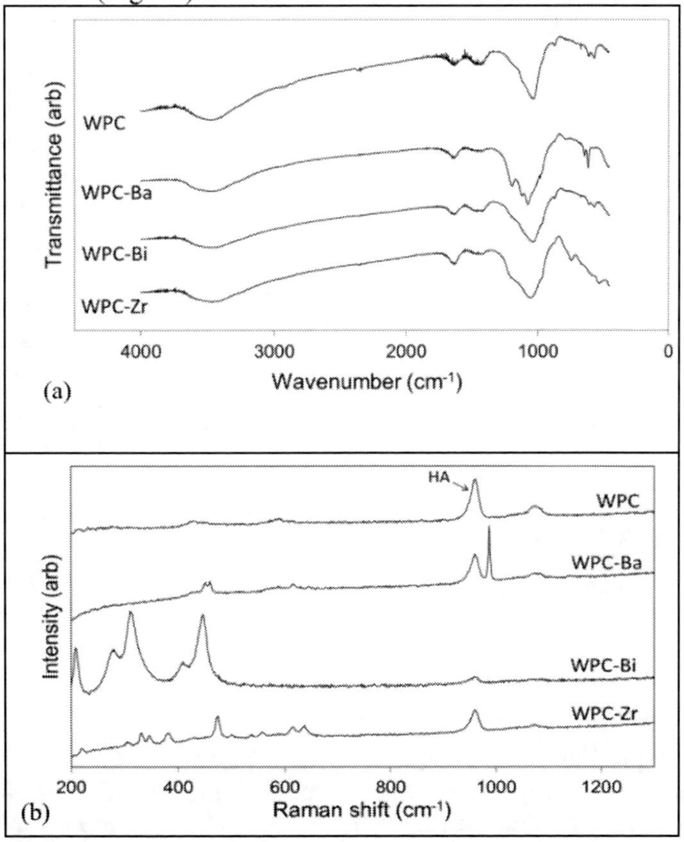

Figure 3. (a) FTIR spectra; and (b) Raman spectra of samples WPC, WPC-Ba, WPC-Bi and WPC-Zr, following immersion in SBF for 7 days at 37 °C.

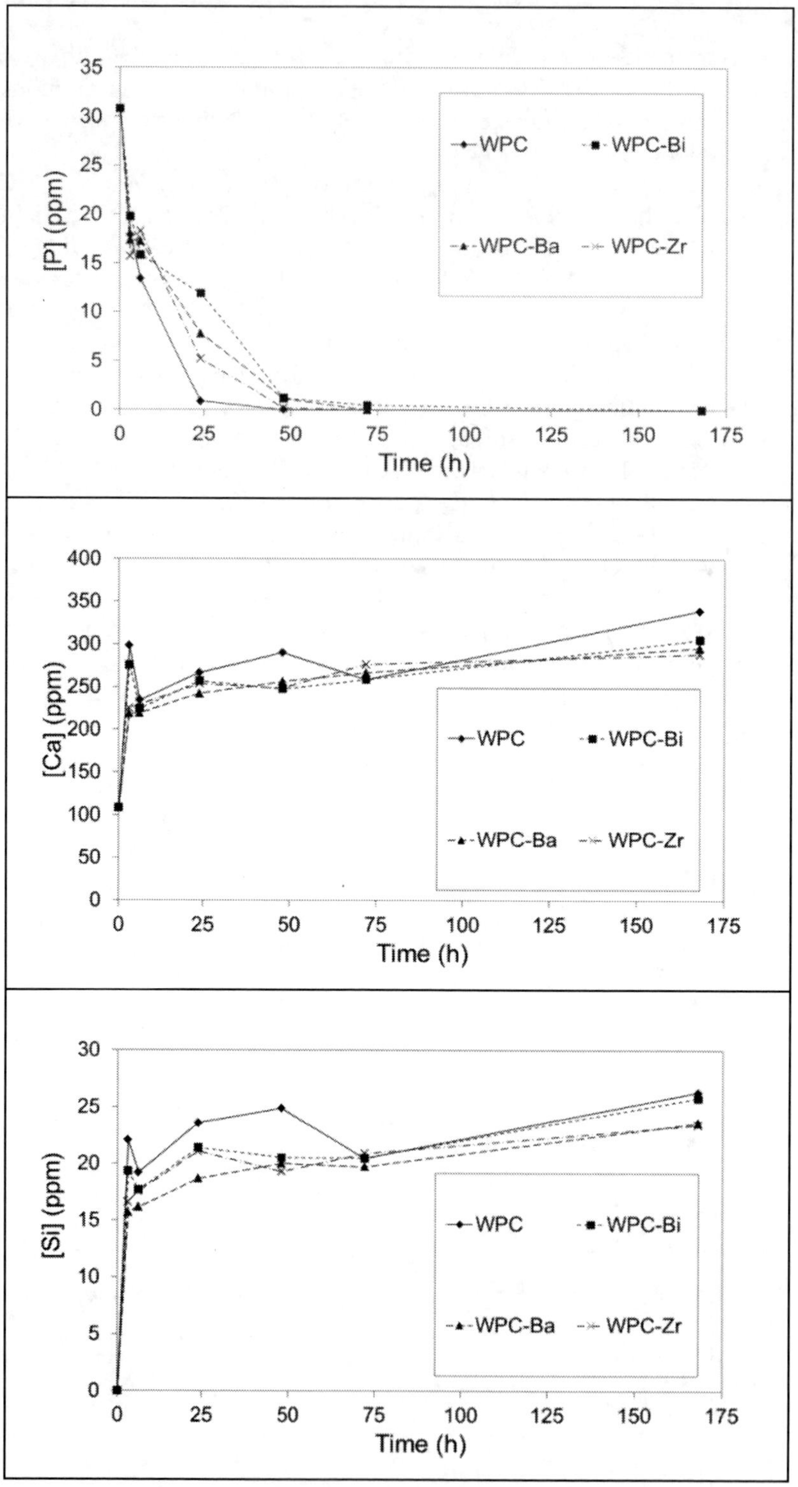

Figure 4. SBF solution profiles for samples WPC, WPC-Ba, WPC-Bi and WPC-Zr.

The concentration profiles of P, Ca and Si in the supernatant SBF solutions in contact with the cement samples are shown in Fig. 4. In all cases, the ongoing reduction of phosphate species in the SBF solution corresponds with the precipitation of HA on the surfaces of the cements and the concurrent release of calcium ions is predominantly attributed to the dissolution of calcium hydroxide from the cement matrix. After 24 hours, the extent of uptake of phosphate ions by the cements was found to be of the following order: WPC > WPC-Zr > WPC-Ba > WPC-Bi, indicating that the initial rate of HA deposition was the slowest for the cement mixed with the proprietary radiopacifier. However, after 1 week, the SBF solutions in contact with all cements were essentially depleted of phosphate ions, indicating that there is little difference among the bioactivities of the different cements.

The observed dissolution profiles of Ca and Si indicate that the incorporation of the radiopacifiers does not compromise the release of these species from the cement matrix. The release of Ca and Si is acknowledged to be relevant to the promotion of new tissue formation, and in this respect all of the experimental radiopacifiers performed similarly [11]. It should be noted, that throughout the duration of the investigation, the concentrations of Al, Ba, Bi and Zr were found to be below the limit of detection (~0.1 ppm) in SBF, which indicates that they are essentially insoluble in these cement matrices.

Recent studies have indicated that a number of candidate radiopacifiers, including barium sulphate and zirconium oxide, may possess superior physicochemical, mechanical or biological properties to those of bismuth oxide [3-8]. Mineral trioxide aggregate is expensive, and accordingly, Portland cement is currently being considered as an alternative root-filling material which could be mixed with a suitable radiopacifier by the clinician [4, 5-7]. In this respect, the findings of this study indicate that barium sulphate and zirconium oxide are superior to bismuth oxide with respect to setting time. This study has also demonstrated that there is no significant difference in the hydroxyapatite-forming ability or Ca- and Si-release among the experimental cements blended with any of the radiopacifiers.

4 Conclusions

This paper reports the comparative impact of 20 wt% of barium sulphate, zirconium oxide or bismuth oxide on the setting times and in vitro bioactivity of white Portland cement. The proprietary radiopacifying agent, bismuth oxide, prolonged the initial and final setting times of the cement; whereas, the two alternative radiopacifiers were found to have no effect on this parameter. The in vitro bioactivity of all cements was confirmed by the formation of hydroxyapatite on their surfaces in simulated body fluid (SBF) within 7 days, and the release of Ca and Si species from the cement matrix was not compromised by the presence of any of the radiopacifying agents. Fourier transform infrared spectroscopy, the traditional technique for the identification of HA in bioactivity studies, was not suitable for the analysis of these cement systems owing to the overlap of signals from each of the radiopacifiers with the characteristic P-O bending modes of HA in the 570 – 610 cm^{-1} region of the spectrum. In this respect, the P-O band at 965 cm^{-1} in the Raman spectrum was found to be a more appropriate means of detection of HA since it is discrete with respect to all other signals arising from the radiopacifying agents and cement phases.

References

1. Q. Li, N.J. Coleman, Dent. Mater. J. **34**, 458 (2015)
2. B.W. Darvell, R.C.T. Wu, Dent. Mater. **27**, 407 (2011)
3. K.S. Coomaraswamy, P.J. Lumley, M.P. Hofmann, J. Endod. **33**, 295 (2007)
4. N. Marques, N. Lourenço Neto, A.P. Fernandes, C. Rodini, M. Hungaro Duarte, D. Rios, M.A. Machado, T. Oliviera, J. Microscopy **260**, 281 (2015)
5. E.A. Bortoluzzi, J.M. Guerreiro-Tanomaru, M. Tanomaru-Filho, M.A.H. Duarte, Oral Surg. Oral Med. Oral Pathol. Oral Radiol. Endod. **108**, 628 (2009)
6. D. Antonijević, D. Ilić, V. Medić, S. Dodić, K. Obradović-Djuričić, Z. Rakočević, Vojnosanitetski Pregled **71**, 1006 (2014)
7. M.A.H. Duarte, G.D.D. El Kadre, R.R. Vivan, J.M. Guerreiro-Tanomaru, M. Tanomaru-Filho, I.G. de Moraes, J. Endod. **35**, 737 (2009)
8. N.J. Coleman, Q. Li, Mater. Sci. Eng. C, **33**, 427, (2013)

9. T. Kokubo, H. Takadama, Biomaterials **27**, 2907 (2006)
10. B. Czarnecka, N.J. Coleman, H. Shaw, J.W. Nicholson, Dent. Med. Probl. **45**, 5 (2008)
11. N.J. Coleman, J.W. Nicholson, K. Awosanya, Cement Concr. Res. **37**, 1518 (2004)
12. P. Taddei, A. Tinti, M.G. Gandolfi, P.L. Rossi, C. Prati, J. Raman Spec. **40**, 1858 (2009)
13. M.G. Gandolfi, K. Van Landuyt, P. Taddei, E. Modena, B. Van Meerbeek, C. Prati, J. Endod. **36**, 851 (2010)
14. S.N. White, Chem. Geol. **259**, 240 (2009)
15. I. Dobrosz-Gómez, M.A. Gómez-García, J. Bojarska, M. Kozanecji, J.M. Rynkowski, Comptes Rendus Chimie **18**, 1094 (2015)
16. N. Garg, K. Wang, S.W. Martin, Cement Concr. Res. **53**, 91 (2013)

 DOI: 10.1051/matecconf/201710903004

A Waste-derived Lithium Metasilicate Basic Catalyst

Victoria Katherine Elmes[a], Andrew Paul Mendham and Nichola Jayne Coleman

Faculty of Engineering and Science, University of Greenwich, Chatham Maritime, Kent, ME4 4TB, UK

Abstract. A mixed product of lithium metasilicate (74%) and minor proportions of calcium carbonate, lithium carbonate and calcium hydroxide was prepared by hydrothermal synthesis from waste green container glass in 4 M lithium hydroxide solution at 125 °C. The reaction product was characterised by powder X-ray diffraction and Fourier transform infrared spectroscopy. The lithium metasilicate product was found to be a successful heterogeneous catalyst for the Knoevenagel synthesis of ethyl trans-α-cyanocinnamate from benzaldehyde and ethyl cyanoacetate.

1 Introduction

Soda-lime-silica glass (SLSG) containers form a large proportion of the urban waste-stream in every developed nation. A number of projects have been carried out to 'upcycle' surplus soda-lime-silica glass into potential value-added materials such as ion-exchangers, lightweight aggregates and ceramics [1-4]. In this respect, a recent study reports the hydrothermal synthesis of lithium metasilicate (Li_2SiO_3) from waste container glass and its potential to act as a carrier for zinc ions [2]. Lithium metasilicate is a synthetic alkaline inosilicate with applications as a breeder material for tritium, ionic conductor, gas-sensor and CO_2-sorbent [5, 6].

Basic inorganic waste-derived heterogeneous catalysts are currently being considered for various organic reactions [7]. These include a wide range of mixed oxide materials from fly ashes, slags and red muds, silica-rich phases from slags and rice husk ash, and calcium-rich phases derived from bones, molluscs and egg shells [7]. In particular, lithium metasilicate, prepared by calcination of rice husk ash and lithium carbonate, has proven successful in the catalysis of the transesterification of soybean oil with methanol to produce biodiesel [8].

The Knoevenagel condensation reaction of an aldehyde or ketone with an activated methylene group is widely employed in the chemical and pharmaceutical industries for the formation of carbon-carbon bonds [9]. This reaction is generally carried out in solution under basic homogeneous catalysis. Environmentally benign solid basic catalysts have been proposed as a superior alternative to the traditional homogeneous catalysis for the Knoevenagel reaction which presents problems associated with the separation of the product and the disposal of the large volume of caustic waste liquor [9].

In this study, a lithium metasilicate-bearing mixed oxide product was prepared hydrothermally from waste SLSG in 4 M lithium hydroxide solution at 125 °C and characterised by powder X-ray diffraction analysis (XRD) and Fourier transform infrared spectroscopy (FTIR). The possibility of

[a] Corresponding author: Vickie.elmes@gmail.com

using this lithium metasilicate product as a basic heterogeneous catalyst for the Knoevenagel reaction between benzaldehyde and ethyl cyanoacetate was explored.

2 Materials and methods

2.1 Preparation and characterisation

Impure lithium metasilicate was prepared from discarded green soda-lime-silica glass (SLSG) bottles by adapting the hydrothermal method described in reference 2. In brief, 3.0g of SLSG and 60 cm^3 of 4M $LiOH_{(aq)}$ were heated at 125 °C in hermetically sealed PTFE reaction vessels under autologous pressure for 14 days. The oxide analysis of the glass was determined by X-ray fluorescence spectroscopy [1, 2] and is given in Table 1. The reaction was carried out in triplicate and the products were washed with deionised water to pH ~ 10 and dried to constant mass in air at 40 °C prior to analysis and use.

Table 1. Oxide analysis of waste green glass.

Oxide component	Mass %
SiO_2	72.15
Na_2O	13.21
CaO	10.48
Al_2O_3	1.48
MgO	0.94
K_2O	0.59
Fe_2O_3	0.46
SO_3	0.28
Cr_2O_3	0.27

The reaction products were analysed by powder XRD using a Bruker D8 diffractometer with Cu Kα = 1.5406 Å, a step size of 0.02° in the 2θ range from 5 to 60° and a measuring time of 1 s per step. X-ray diffraction data were compared with JCPDS files using DIFFRAC.EVA software (Bruker, Germany). Quantitative analysis was carried out by Rietveld refinement using a fundamental parameters approach with full axial model using TOPAS 4.2 software (Bruker, Germany) [10]. The lithium metasilicate structure used for the quantitative analysis was described by Tang et al. [11] and all other structures were selected from the TOPAS 4.2 structure database (Bruker, Germany). The FTIR spectrum of the reaction products was acquired using a Perkin Elmer Spectrum Two spectrometer between 450 and 4000 cm^{-1} wavenumbers, with 16 scans at a resolution of 4 cm^{-1}.

2.2 The Knoevenagel condensation reaction

The catalytic properties of lithium metasilicate were investigated using the Knoevenagel condensation reaction between benzaldehyde and ethyl cyanoacetate to produce ethyl trans-α-cyanocinnamate [9, 12]. 0.25 g of lithium metasilicate were added to a mixture of benzaldehyde (19.6 mmol) and ethyl cyanoacetate (18.8 mmol) in 50 cm^3 of ethanol, in a three-necked round bottom flask, equipped with a reflux condenser, thermometer and magnetic stirrer. The mixture was heated at 80 °C for 1.5 h, under constant stirring. At time intervals of 0, 15, 60 and 105 mins, 1 cm^3 aliquots of the reaction mixture were withdrawn for analysis via gas chromatography (GC), using an Agilent 7683B instrument equipped with an HP-1 capillary column and a flame ionisation detector (FID). The same reaction was conducted in the absence of the lithium metasilicate product. All reactions were carried out in triplicate. The ethyl trans-α-cyanocinnamate product was also confirmed by gas chromatography with mass spectrometry (Perkin Elmer turbomass DB5 spectrometer).

3 Results and discussion

3.1 Characterisation

The powder XRD pattern of the hydrothermal reaction product of waste green container glass and 4 M aqueous lithium hydroxide solution at 125 °C for 14 days is shown in Fig. 1. These data confirm that the principal reaction product is orthorhombic lithium metasilicate (JCPDS file 29-0828), and that calcite, lithium carbonate and calcium hydroxide, are also present in minor proportions. The relative proportions of the components of the lithium metasilicate product are listed in Table 2. The weighted profile R-factor (Rwp), for the refinement was 8.15, which demonstrates that the fittings are satisfactory.

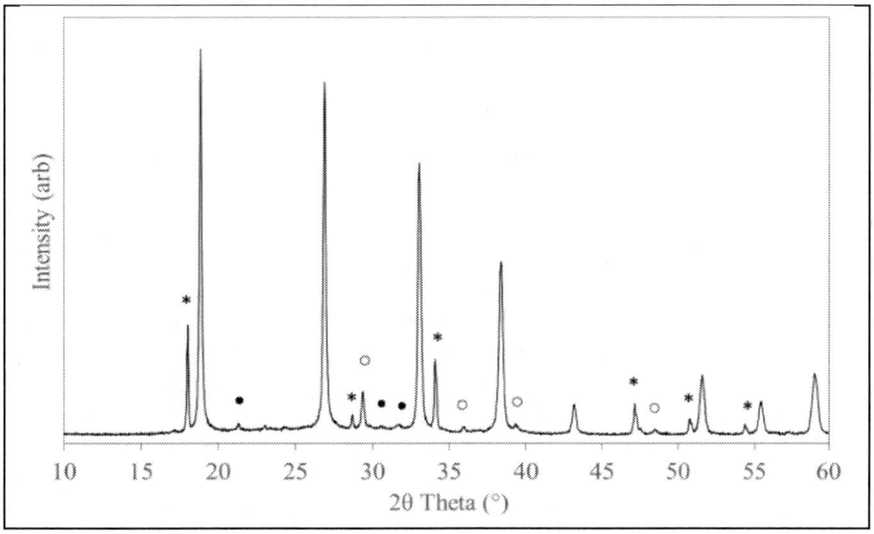

Figure 1. XRD pattern for the hydrothermal reaction product (Key: ● Li_2CO_3; ○ $CaCO_3$; * $Ca(OH)_2$).

Table 2. Phase composition of lithium metasilicate product.

Phase	Mass (%)
Li_2SiO_3	73.52
Amorphous phase	17.05
$CaCO_3$	2.35
Li_2CO_3	2.74
$Ca(OH)_2$	4.34

The composition of the impure lithium metasilicate obtained in this study is similar to that reported previously for a hydrothermal reaction product of colourless SLSG synthesised in 4 M lithium hydroxide at 100 °C for 5 days [2]. The increase in reaction temperature and time to 125 °C and 14 days, respectively, is seen to increase the proportion of lithium metasilicate from 67.9 to 73.5%, and to eliminate the minor reaction product, tobermorite.

 DOI: 10.1051/matecconf/201710903004

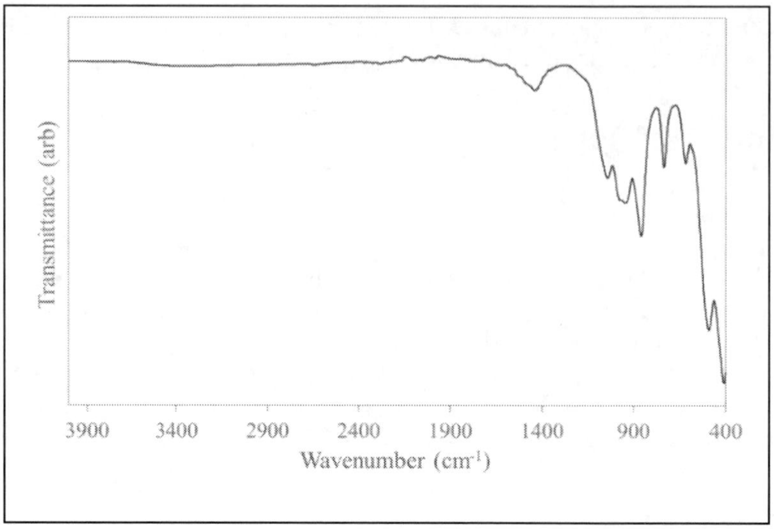

Figure 2. FTIR spectrum for the hydrothermal reaction product.

The FTIR spectrum of the lithium metasilicate reaction product is presented in Fig. 2 and closely resembles those of lithium metasilicate reported in the literature [2, 13]. Si-O-Si vibrations give rise to bands at 1045, 984, 736 and 619 cm^{-1}. The signal at 523 cm^{-1} is attributed to Si-O-Li deformations and O-Si-O vibrations give rise to the bands at 945 and 863 cm^{-1}.

3.2 The Knoevenagel condensation reaction

The potential ability of the lithium metasilicate product to catalyse the Knoevenagel condensation was tested on the reaction of benzaldehyde and ethyl cyanoacetate at 80 °C to produce ethyl trans-α-cyanocinnamate. The ethyl trans-α-cyanocinnamate product was confirmed by GC-MS, and the reaction, in the presence and absence of the catalyst, was monitored by GC. The relative concentrations of the starting reagents were used to evaluate the rate of reaction. The solution profiles of benzaldehyde and ethyl cyanoacetate as functions of time are plotted in Figure 3. These data indicate that the rate of reaction is significantly enhanced in the presence of the heterogeneous lithium metasilicate catalyst. It should be noted, that two additional by-product phases, both of relative peak intensity below 0.1%, were detected by GC. These products have yet to be identified by mass spectrometry.

A previous study by Chen et al. [8], has also confirmed that lithium metasilicate derived from waste rice husk ash has been used to successfully catalyse the production of biodiesel from soybean oil with methanol. In this respect, waste-derived lithium metasilicate appears to be a promising basic heterogeneous catalyst for organic synthesis.

An investigation of the recovery and reuse potential of the SLSG-derived lithium metasilicate product for the catalysis of the Knoevenagel condensation and its ability to catalyse other organic reactions is now underway. The next stage of this research will also consider the economic feasibility of scaling-up this system.

4 Conclusions

This study has confirmed that a mixed product of lithium metasilicate (74%) with minor proportions of calcium carbonate, lithium carbonate and calcium hydroxide can be prepared by hydrothermal reaction of waste green container glass in 4 M lithium hydroxide solution at 125 °C for 14 days. The lithium metasilicate product was found to successfully catalyse the Knoevenagel synthesis of ethyl trans-α-cyanocinnamate from benzaldehyde and ethyl cyanoacetate. These findings indicate that this material is a promising basic heterogeneous catalyst for organic synthesis.

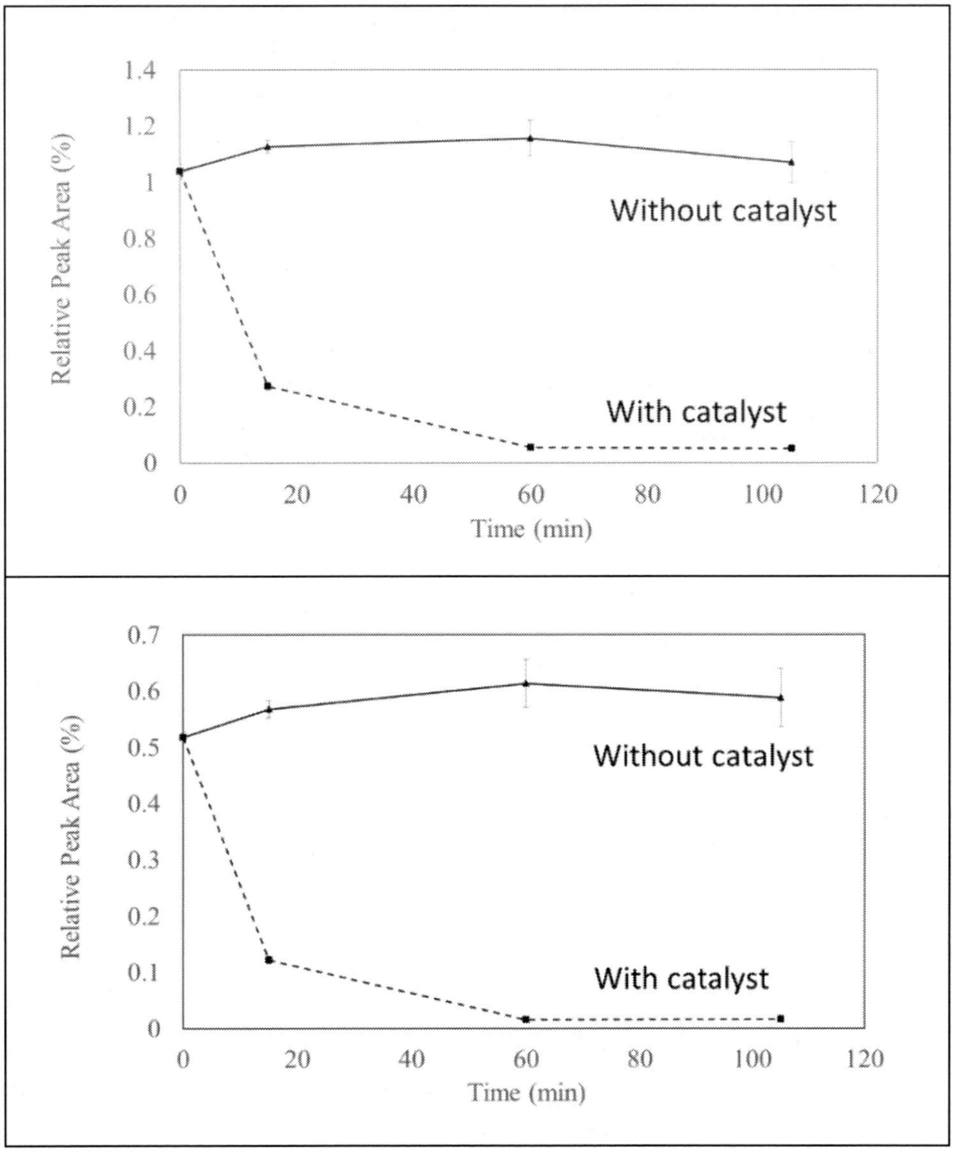

Figure 3. Concentration profiles of benzaldehyde (top) and ethyl cyanoacetate (bottom).

References

1. N.J. Coleman, Q. Li, A. Raza, Physicochem. Probl. Miner. Process. **50**, 5 (2014)
2. N.J. Coleman, A.P. Hurt, A. Raza, Physicochem. Probl. Miner. Process. **51**, 685 (2015)
3. F. Matteucci, M. Dondi, G. Guarini, Ceram. Int. **28**, 873, (2002)
4. C.A. Velis, C. Franco-Salinas, C. O'Sullivan J. Najorka, A.R. Boccaccini, C.R. Cheeseman, Environ. Sci. Technol. **48**, 7527 (2014)
5. H. Pfeifer, P. Bosch, S. Bulbulian, J. Nucl. Mater. **257**, 309 (1998)
6. J. Ortiz-Landeros, M.E. Contreras-Garcia, C. Gomez-Yanez, H. Pfeifer, J. Solid State Chem. **184**, 1304 (2011)
7. J.A. Bennett, K. Wilson, A.F. Lee, J. Mater. Chem. A **4**, 3617 (2016)

8. K.-T. Chen, J.-X. Wang, Y.-M. Dai, P.-H. Wang, C.-Y. Liou, C.-W. Nien, J.-S. Wu, C.-C. Chen, J. Taiwan Inst. Chem. Eng. **44**, 622 (2013)
9. N.P. Tangale, S.K. Sonar, P.S. Niphadkar, P.N. Joshi, J. Ind. Eng. Chem. **40**, 128 (2016)
10. R.W. Cheary, A.A. Coelho, J. Appl. Cryst. **25**, 109 (1992)
11. T. Tang, P. Chen, W. Luo, D. Luo, Y. Wang, J. Nucl. Mater. **420**, 31 (2012)
12. D. Jain, M. Mishra, A. Rani, Fuel Process. Technol. **95**, 119 (2012)
13. B. Zhang, M. Nieuwoudt, A.J. Easteal, J. Am. Ceram. Soc. **91**, 1927 (2008)

One-pot Synthesis of Octyne-Ruthenium on Carbon Nanoparticles

Kanda Wongwailikhit[1,a], Rawewan Chaisinchareon[2] and Shaowei Chen[2]

[1]*Department of Chemistry, Faculty of Science, Rangsit University,Patumtani,Thailand*
[2]*Department of Chemistry and Biochemistry, University of California, Santa Cruz, US*

Abstract. The attempts to manipulate ruthenium nanoparticle by the passivation of π bonds linkage is of interest for many years. That is the way to enhance its optical properties and fluorescence characteristics which can promote the usage for sensor application. Other view, the usage of carbon nanoparticle is governed in many aspects including its fluorescence properties. Therefore, the combination between those two valued nanoparticles was set by conducting the simple synthesis method. With the as-prepared carbon nanoparticles, all other reagents (ruthenium (III) chloride, octyne and Sodium borohydride) were mixed in the same batch. The ratio of carbon substrate, ruthenium (III) chloride and octyne was 10: 1: 3. The particle yielded was then purified and subjected to characterize using some spectroscopy techniques including photoluminescence. The results showed that size of carbon particle before and after ruthenium deposition were 5.0 and 6.3 nanometers, respectively. Octyne was coordinated self-assembly on the ruthenium surface which was 8.1 nanometers in diameter. Moreover, octyne-protected ruthenium on carbon nanoparticles showed the remarkably increasing of fluorescence Intensity. Therefore, the functionalization of carbon nanoparticle with octyne-ruthenium can be a promising strategy to develop a novel complex of ruthenium.

1 Introduction

Recently, many reports have been generally successfully prepared alkyl vinylidene –linkage to ruthenium core using the special coordinating ligand for example terminal alkyne [1-5] or some vinyl aromatic [6-7]. It is confirmed that alkyne fragments around the metal core do not only functions as protecting ligands promoting the stability of the nanoparticles but also enhances some optical and fluorescence characteristic via the pi-bond interaction between methylene and metal core[8]. It should be noted that alkynes (and alkenes as well) have been used rather extensively as coordinating ligands in the preparation of varied transition metal complexes [9].With two π bonds, alkynes can form stable complexes in which they bridge two metal centers. Then alkyne can donate total four electrons which support the exploitation of metal−carbon covalent linkages for nanoparticle surface functionalization [10]. The conjugated interfacial bonds can extend the intraparticle charge delocalization between particle-bound functional moieties, leading to increase optical and electronic properties [11].

[a] Corresponding author: kanda@rsu.ac.th

This work, we try to develop the idea of using carbon nanoparticles (CPNs) as the working core instead of pure ruthenium nanoparticle. It is well known that carbon nanoparticle is enriched with carbonyl group which occurred simultaneously with the formation of the CNPs, and effective surface states for luminescence such and used as chemical sensing, biological labeling, and optoelectronics [12]. Herein, the ruthenium metal was attempted to deposit onto CPNs surface and then covered the outer-shell with alkyne. The process was conducted in one-pot synthesis and testified whether the deposition was be succeeded. The properties of the products were determined with some spectrophotometric methods including the photoluminescence properties.

CPNs was prepared from soot with the same direction reported previously [13]. The nanosized carbon particles synthesized composed of sp^2 graphitic carbons in the crystalline core with good photoluminescence in the visible range. Moreover, it was reported that and the surface of carbon was plenty of oxygenated functional moieties. Ruthenium chloride and sodium borohydride were used as source of metal ion and reducing agent, respectively. Octyne was chosen to be the functionalized ligand onto ruthenium surface.

2 Chemicals

Ruthenium chloride ($RuCl_3$, 99+%, ACROS), 1-octyne (Alfa Aesar, 98%), Nitric acid (HNO_3, 69.8%, Fisher), sodium carbonate (Na_2CO_3, 99+%, Aldrich), methanol (CH_3OH, Fisher Scientific] and sodium borohydride ($NaBH_4$, 99%, ACROS) were all used as received. Water was supplied by a Barnstead Nanopure Water System (18.3 MΩ 3 cm).

3 Methodologies

3.1 Synthesis of Carbon nanoparticles

Carbon Nanoparticles was prepared in the same procedure of published articles [13-14]. Briefly described as, Carbon soot was collected on the inside wall of a glass beaker by placing the beaker upside-down above the flame of a natural gas burner. After being washed with acetone and dried at 110°C, 100 mg of the soot was then refluxed in 10 mL of 5 mol/L HNO_3 at 140°C for 12 h. After the mixture was cooled to room temperature, the brownish solution was neutralized by Na_2CO_3 and loaded into dialysis bags (cutoff molecular weight 8000–14,000) for dialysis against ultra-pure water for three days to purify the CNPs. Then CNPs was dialyzed against ultra-pure water for 3 days. Water in dialysis process was replenished every day. CNPs were collected by centrifugation at 16,000 rpm. The precipitation was dried under nitrogen flow and dispersed again in ultra-pure water. Big particles were removed using ultrafiltration till brown solution of CNPs was obtained. This solution was kept at 4°C until use.

3.2 Synthesis of Alkyne-Ruthenium on Carbon nanoparticles

Octyne-ruthenium was deposited on CPNs by mixing $RuCl_3$, $NaBH_4$, and CNPs at the ratio of 3:3:10 in pure water. Briefly describe, 10 mg of CNPs was mixed with 1 mg $RuCl_3$ in 10 mL of water and stir over night. Octyne was added at the mole ratio between $RuCl_3$ and octyne of 1:3. Then, a calculated amount of $NaHB_4$ was added in a slow drop-wise fashion. The solution color changed gradually from light brown to dark brown, signifying the formation of Ruthenium under reduction reaction. The resulting solution was allowed to stir at room temperature over 3 h. After the reaction, the carbon-supported Ruthenium nanoparticles were treated with methanol and separate over the ultracentrifuge. The particles were found to be readily soluble in typical non-polar solvents such as toluene, tetrahydrofuran, chloroform, and dichloromethane, and not soluble in polar solvents such as alcohols, acetone, and acetonitrile. Spectroscopic measurements were then carried out to examine the product. A set of experiment was also run with the same procedure except adding octyne. The resulting particle was used to compare the result between octyne-protected ruthenium on carbon nanoparticle and bare ruthenium on carbon nanoparticle

4 Results

As mention above, aim of our work was to prove if the synthesis process was able to short with one step of relevant mixtures. The target yield was the nanoparticle of ruthenium metal deposited on CPNs (denoted as C/Ru) and octyne was linked on top of ruthenium surface with the π-covalent bond (denoted as C/Ru/Oct). Below implied the results which illustrating the success of the production.

The formation of C/Ru/Oct nanoparticles was first verified by UV-visible analysis measurements. Figure 1 (a) shows the UV–visible absorption spectrum of carbon nanoparticles in water (0.1 M). The particles exhibited a major absorption peak at around 200 nm ascribing to the p–p* electronic transitions of internal (sp^2) graphitic carbons. Figure 1 (b) responses the changes of carbon nanoparticle spectrum to the featureless exponential decay profile anticipating for nanosized Ru particles [15]. This suggested the success of the ruthenium deposition. Figure 1(c) depicts a UV-vis spectrum of the C/Ru/Oct nanoparticles prepared. It can be seen that the Ru nanoparticles were all well dispersed without apparent aggregation, suggesting effective passivation of the alkyne ligands on the Ru particle surfaces.

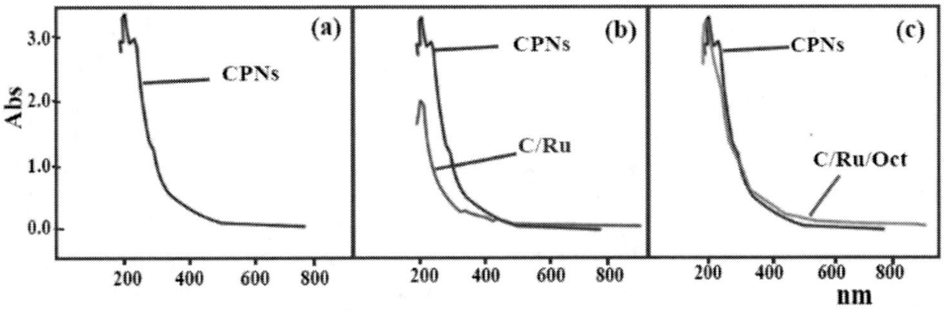

Figure 1. The UV–visible absorption spectrum of products in water (0.1 M), (a) CPNs (b) C/Ru (c) C/Ru/Oct

All particles were investigated using particle size analysis and the results were shown in figure 2. It was corresponding to the results interpreting by UV-visible spectrum. Figure 2 (a) showed the average of particle size of CPNs about 5.0 nm. After the ruthenium deposition, C/Ru, size of particle was raised to about 6.3 nm (figure 2 (b)). Finally with the passivation of octyne, C/Ru/Oct, size of average particle was increased to 8.1 nm (figure 2 (c)) which suggesting the effective passivation of the octyne ligands on the ruthenium surfaces. This suggested that thickness of ruthenium layer plus octyne layer was about 3.1 nm. It was relatively close to the octyne passivated on ruthenium nanoparticle synthesized by Chen, W. et. al. [16] which was 2-3 nm sized average. Herein, it was reasonable to conclude the successful production of our process.

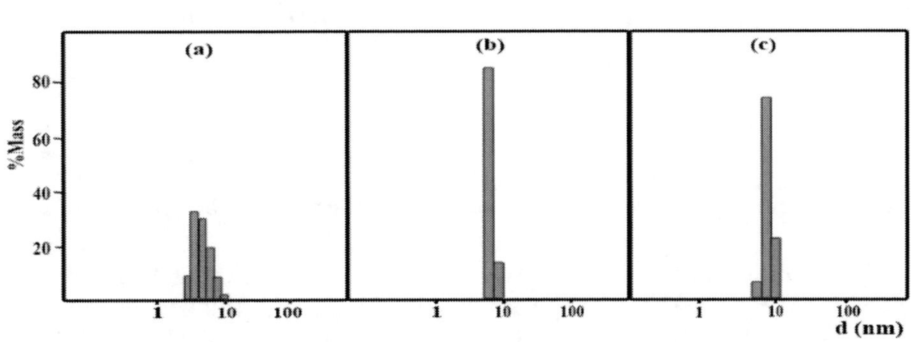

Figure 2. Particle size analysis results obtained from (a) CPNs (b) C/Ru (c) C/Ru/Oct

The structure of pure carbon nanoparticles and C/Ru/Oct were then examined by FTIR spectrum and compare with CPNs as shown in figure 3.

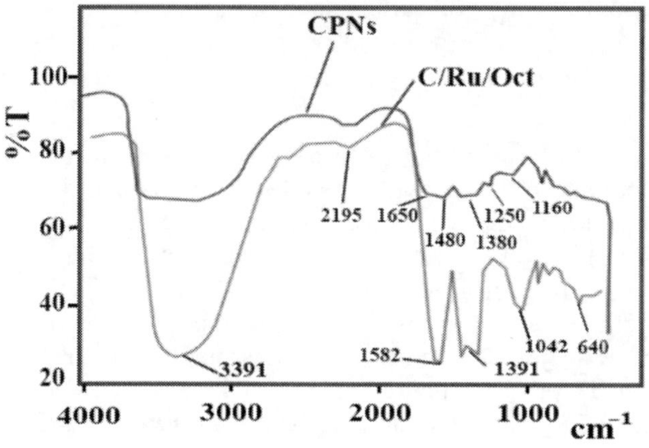

Figure 3. FTIR spectra of CPNs and C/Ru/Oct

For the CPNs, a characteristic peak at 1160cm^{-1} was assigned to C–C bonds. Also, the spectrum showed the carbonyl characteristic broad peak at 1650 cm^{-1}, which was assigned to the carbonyl group from quinine or ring structure. [17] Moreover, at 1480 cm^{-1} suggesting –C-H and 1380 cm^{-1} showed the –C-O- and 1250 cm^{-1} showed the O=C-OH which was the special characteristic of carbon nanoparticles synthesized using soot [18].

For C/Ru/Oct, a characteristic peak at 1582 cm^{-1} was assigned to C=C bond. This should be said that C=C band was such strong and showed the plenty of C=C bond after passivation with Octyne. The broad band centered around 3500 cm^{-1} most likely arose from residual water. The —C≡C— vibrational stretch of alkynes was shown at 2195 cm^{-1} which typically observed for monomeric alkynes as a weak band within 2260 cm^{-1}[19]. This shift was responsible to the complex between metal and alkyne [20]. The absent of peaks at 1250 cm^{-1} suggested that O=C-OH was reduced in the synthesis process and form R-O-H or R-O-R which was absorbed at 1042 cm^{-1}. The strong peak at 1390 cm^{-1} showed the increase up of –C-H bond from octyne tail [21]. This suggested that the octynyl fragments were indeed bound onto the metal surface.

Further supporting evidence was manifested in the photoluminescence measurements and shown in figure 4.

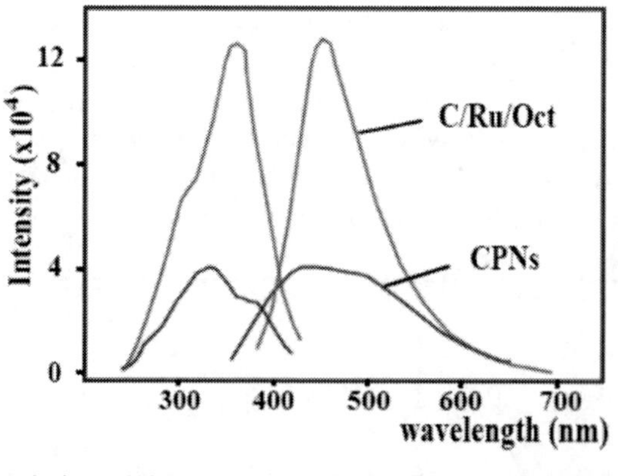

Figure 4. Excitation (left) and emission (right) spectra of CPNs (black) and C/Ru/Oct (red) at 25 °C. The particle concentration was 0.1 mg/mL.

Figure 4 showed the excitation and emission spectra of the corresponding carbon nanoparticles before and after octyne-protected Ruthenium treatments in. It illustrated that the excitation and emission peak energies had practically changed from λex = 330 nm and λem = 430-500 nm (440) to

λex = 360 nm and λem = 450 nm. This suggested a different radiative decay pathway for the photoexcited electrons in the original CPNs and treated particles. In sharp contrast, the photoluminescence intensity exhibited drastic enhancement of more than the original ones. This significant improvement in the emission intensity was most likely due to the success of the deposition of ruthenium and the passivation of octyne over on ruthenium. Additionally, it can be seen that the adsorption peaks became narrower and contributed to the small red shift after treatment on metal nanoparticles [22]. The well-defined excitation and emission peaks shown at 360 and 450 nm, respectively were corresponding to the previous studies of ruthenium nanoparticles passivated by 1-dodecyne [7] and 1-octyne [16]. The high intensity of photoluminescence observed with the C/Ru/Oct nanoparticles strongly suggests that intraparticle charge delocalization took place as a result of the strong Ru=C interfacial bonding interactions. Again, the successful incorporation of 1-octyne onto the nanoparticle surface further confirms the formation of a Ru−vinylidene interfacial linkage with nanoparticles.

5 Conclusions

The attempt to use CPNs as the center core for depositing ruthenium metal and the outer surface was protected with octyne was successful with a very simple hydrothermal method. CPNs was prepared from soot and mixed with octyne, $RuCl_3$ and sodium borohydride as the reducing agent. The as-prepared product was collected, purified and subjected to characterize some properties. The results shown that the one pot synthesis could produce the novel nanoparticles having the carbon nanoparticle core with ruthenium plated on the surface and the outer most shell was the vinylidene linkage of terminal octyne. Moreover, because of the specific photoluminescence properties, the nano-complex of carbon-ruthenium protected with octyne could be a promising particle for supporting the novel sensor technology.

References

1. X. Kang, S. Chen, Nanoscale. **4**, 4183 (2012)
2. H. Sung, T. Her, W. Su, C. Cheng, Molecules. **17**, 8533 (2012)
3. N.T. Lucas, M.P. Cifuentes, L.T. Nguyen, M.G. Humphrey, Cluster Sci. **12**, 201 (2001)
4. C. Chen, H. Tong, C. Hsu, C. Lee, Y. Fong, Y. Chuang, Y. Lo, Y. Lin, Y. Wang, Organometallics. **28**, 3358 (2009)
5. H. Kopf, C. Pietraszuk, E. Hübner, N. Burzlaff, Organometallics. **25**, 2533 (2006)
6. V. Dragutan, I. Dragutan, Platinum Metals Rev. **48**, 148 (2004)
7. X. Kang, N.B. Zuckerman, J.P. Konopelski, S. Chen, J. Am. Chem. Soc. **134**, 1412 (2002)
8. W. Chen, S. Chen, F. Ding, H. Wang, L.E. Brown, J.P. Konopelski, J. Am. Chem. Soc. **130**, 12156 (2008)
9. Structure and bonding of alkyne complex. Available at http://www.ilpi.com/organomet/alkyne.html
10. W. Chen, L.E. Brown, J.P. Konopelski, S. Chen, Chem. Phys. Lett. **471**, 283 (2009)
11. W. Chen, N.B. Zuckerman, J.W. Lewis, J.P. Konopelski, S. Chen, J. Phys. Chem. C.**113**, 16988 (2009)
12. M.S. Dresselhaus, G. Dresselhaus, Nanostruct. Mater. **9**, 33 (1997)
13. B.R. Selvi, D. Jagadeesan, B.S. Suma, G. Nagashankar, M. Arif, K. Balasubramanyam, M. Eswaramoorthy, T.K. Kundu, Nano Lett. **8**, 3182 (2008)
14. L. Tian, D. Ghosh, W. Chen, S. Pradhan, X. Chang, S. Chen, Chem. Mater. **21**, 2803 (2009).
15. X. Kang, S. Chen, Nanoscale. **4**, 4183 (2012)
16. W. Chen, N.B. Zuckerman, X. Kang, D. Ghosh, J.P. Konopelski, S. Chen, J. Phys. Chem. *C*,**114**, 18146 (2010)
17. Z. Yunusa, M.N. Hamidon, A. Ismail, M.M. Isa, M.H. Yaacob, S. Rahmanian, S.A. Ibrahim, A.A. Shabaneh, Sensors. **2015**, 15, (2015)

18. X. Sun, J. He, Y. Meng, L. Zhang, S. Zhang, X. Ma, S. Dey, J. Zhaoc, Y. Lei, J. Mater. Chem. A, 4, 4161 (2014)
19. Infrared spectroscopy table. Available at; http://www.ochemonline.com/Infrared_spectroscopy_absorption_table
20. N.J. Long, C.K. Williums, Angewandte Chemie. **42**, 2586 (2003)
21. X. Kang, N.B. Zunkerman, J.P. Konopelski, S. Chen, J.Am.Chem. Soc. **134**, 1412 (2012)
22. Chen, S. Pradhanx, S. Chen, Nanoscale. **3**, 2294 (2011)

Polymer Network Gel Method for Surperfine AlN Powder Preparation

Liuyan Wang [1,a]

[1] *College of architecture and civil engineering, Shenyang University, Shenyang 110000,China*

Abstract. To research mixing uniformity of precursor was helpful to improve the quality of the powder late processing. Layered problems of the precursor comprised of aluminium source and of carbon source were well solved by the polymer network method. High quality raw precursor was synthesized by new process which contained polyacrylamide-gel method and carbothermal process. The synthesized precursor was investigated to obtain porous structure precursor. The results showed that sucrose was more suitable for the preparation of AlN precursor. The effect of reaction temperature and heating time on the phase composition of AlN powder were also studied. In this work, ultrafine nano-sized AlN powder was prepared at low temperature.

1 Introduction

There are many scholars devoted to the low temperature preparation of density AlN ceramic, in order to exert its features such as high thermal conductivity, good electrical resistivity, high melting temperature, good thermal shock resistance, wide band gap, low dielectric constant and high mechanical strength, etc, which may be useful for high integration of semiconductor substrate and electronics packaging applications. High quality raw material (AlN precursor) is believed to be one of the ways to achieve this purpose. For this, we improved the traditional preparation technology, and studied the effect of improvement process on the preparation of ultrafine aluminium nitride.

At present, it was shown that cabothermal reduction method [1-5], high energy ball-milling method [6, 7], mechanical activated method [8, 9], sol-gel technique [10-13], direct-nitridation[14,15] and chemical vapour deposition[16-19], were widely used for the preparation of AlN. However, these synthesis processes suffering from various problems, including inferior crystal structure, non-uniform microstructure and complex process as well as high cost of raw material. So improving the uniformity of precursor, technological process and recovery rate become the key technologies which need to resolve during the preparation of AlN powder.

In this work, C_3H_5NO was used as monomer.$C_7H_{10}N_2O_2$ was used as crosslinking network. $(NH_4)_2S_2O_8$ was used as initiator. And Al $(NO_3)_3 \cdot 9H_2O$ was used as Alumium source for its lower prices.The precursor was prepared by polyacrylamide gel method, which was best solved the layered phenomenon of Al and C source. The polymer gel network was produced by functional groups activation of the double bond using acrylamide as monomer. The activity of ion was reduced by the effect of gel network blocking. And the reunion opportunity in the process of precipitation and

[a] Corresponding author: w.ly428@163.com

sintering was also decreased. These results would lay the foundation for the further studies on the ultrafine of AlN powder.

2 Experimental

2.1 Technological process of AlN precursor

A certain amount of monomer, network agent and initiator (weight ratio of 5:1:1) was added into 0.3 mol/l of aluminum nitrate solution, and was magnetically stirred for 30min.The temperature in this titration process was controlled below 0°C. Meanwhile, carbon sources (sucrose or active carbon) were prepared. Here the molar ratio of C and Al was 6:1. After 3.42 g of sucrose was added into the mixture and stirring for 60min. Then ammonia solution was dripped into the mixtures with the concentration ratio of 1:2 and 3 - 5 ml/min dripping speed. When titration was finished, the reaction mixture was continued to stir for 120min. The prepared wet gel was obtained until the mixture was aggregated at 65°C for 120min. Final AlN precursor was obtained while the wet-gel directly drying at 85°C for 120min in the vacuum drying oven.

All raw materials ($Al(NO_3)_3 \cdot 9H_2O$, $NH_3 \cdot H_2O$, $C_{12}H_{22}O_{11}$, C_3H_5NO, $C_7H_{10}N_2O_2$ and $(NH_4)_2S_2O_8$) used in the preparation of AlN precursor were 99% in purity.

The synthesis of nanosized AlN powder was carried out in vacuum carbothermal reducing furnace. The AlN precursor was heated in a flowing nitrogen gas at various temperatures in the range of 1100-1500°C for 120min.And the flow rate of nitrogen was 10 l/min. Residual carbon in each nitridation product was removed by firing in air at 640°C for 240min[20].

2.2 Characterization

Phase analysis of the nitridation reaction products for AlN precursor calcined at different temperature were carried out by X-ray diffraction (XRD, PhilipPw3040/60) using monochromatic Cu Kα radiation with the X-ray wavelength of 0.154nm, operated at 20kV.

Microstructures of the reaction products were observed using field emission scanning electron microscopy (SEM, SSX-550, Shimadzu Corporation). Element distributions of sample were analyzed by energy dispersive X-ray spectroscopy (EDS). Here EDS was used to assist the SEM analysis.

3 Results and Discussion

3.1 Effect of different carbon sources on the precursor

AlN precursors were synthesized by means of polyacrylamide gel method with ammonia as precipitant, sucrose and activated carbon as carbon source, respectively. Here the molar ratio of C to Al was 2:1.The SEM images of AlN precursor synthesized by different carbon source were shown in Fig.1. It was evident that the section of AlN precursor which used activated carbon as carbon source presents obvious layers. However, AlN precursor with sucrose that contains a homogeneous mixture of very fine alumina and carbon particle can be obtained in this work. Compared with active carbon, sucrose was more suitable for well-proportioned precursor. Sucrose was used as carbon source based on the above reason.

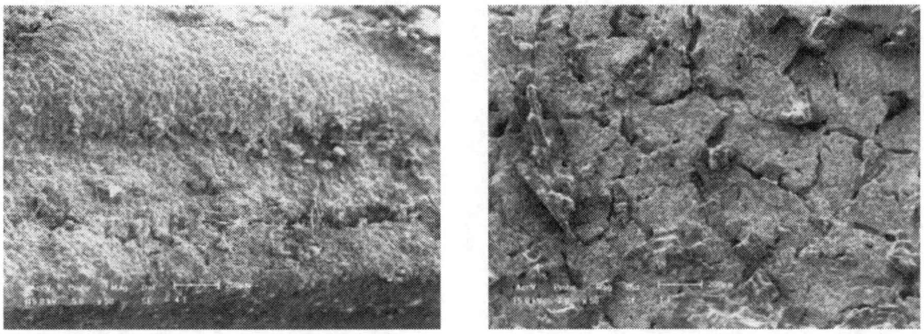

(a) (b)

Figure 1. SEM images of AlN precursor prepared by ammonium hydroxide with different carbon source (a) sucrose (b) activated carbon

3.2 Effect of sucrose on the uniform precursor

The SEM images of AlN precursor under different magnification were shown in Fig.2. The sample was observed under low magnification. And there was no obvious stratification in it (see Fig.2(a).

To further study the microstructure and distribution of precursor sample, it was magnified 2000 times by SEM, as shown in Fig.2(b). Meanwhile, the microstructure of the precursor was porous structure. It would suggest that adding liquid sucrose may accelerate the formation of network structure. The elementary and crosslinking reaction of polymeric hydrogel and small molecule of liquid sucrose can explain this phenomenon very well.

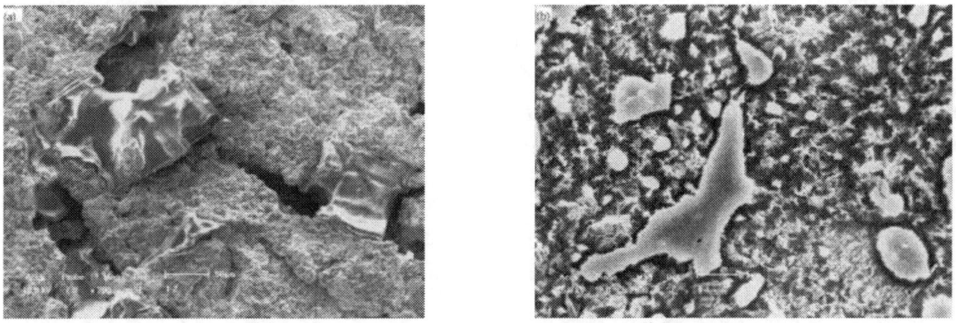

(a) (b)

Figure 2. SEM images of AlN precursor under different magnification (a) 200times magnification (b) 2000times magnification

Fig.3 shows surface scanning images of AlN precursor using ammonium hydroxide and sucrose as raw material. Fig.3a shows that AlN gel precursor has crosslinked and the network structure formed. As shown in Fig.3b, c and d, elements composition and distribution in AlN precursor were also studied by lateral scanning. Fig.3b, c and d show the distribution of Al, C and O atoms in aluminum nitride precursor, respectively. Continuous and uniform distribution of C atoms in AlN precursor was clearly seen from Fig.3c.

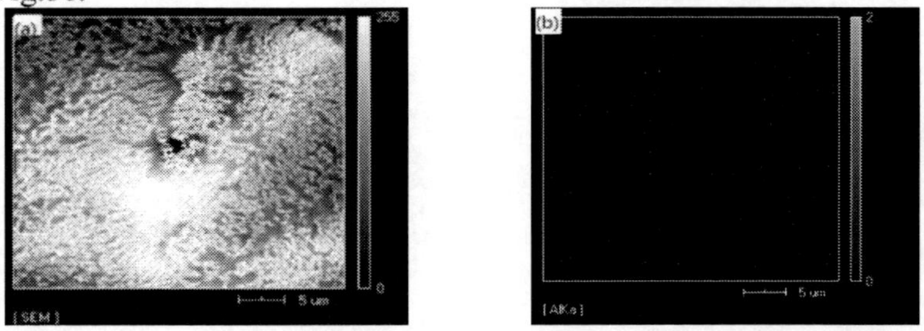

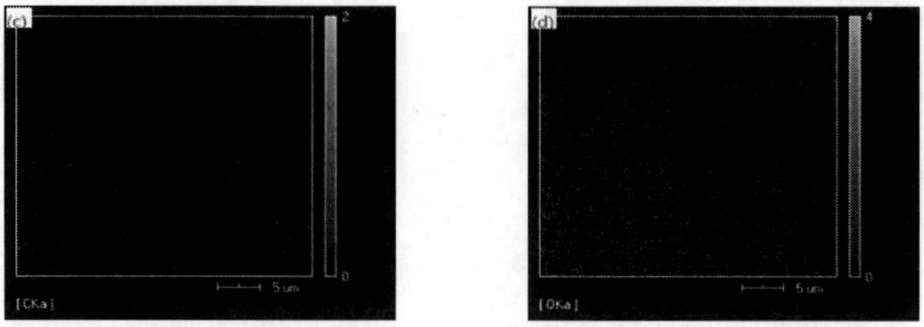

Figure 3. Surface scanning images of AlN precursor

3.3 Effect of reaction temperature on the phase composition of AlN powder

So far, detailed reaction mechanism for the synthesis of AlN powder by carbon thermal reduction method was unclear. Based on some research [21], the nitridation reaction of AlN powders prepared by carbon thermal reduction may be suggested as the following equation:

$$Al_2O_{3(s)} + 3C_{(s)} + N_{2(g)} \rightarrow 2AlN_{(s)} + 3CO_{(g)} \quad (1)$$

The Gibbs free-energy values for reaction were calculated to confirm a driving force. According to some theories in some paper [22], the initial reaction temperature is 1400°C. A series of AlN powders were prepared at various temperatures by the carbothermal reduction method using AlN precursor as raw material.

Fig.4 shows the reaction products of AlN precursor under various temperatures. The XRD patterns of AlN precursor and products that were calcined at different temperature for 2h were shown in Fig.4.

From Fig.4a, the XRD pattern of the sample obtained at 1400°C showed that AlN phase was formed together with the diffraction peaks of Al_2O_3 and C. This suggests that the AlN precursor was not completely sintered.

Compared with JCPDS Card No.25-1133, some diffraction peaks represent the existence of AlN. Fig.4b shows that the intensities of AlN peaks increased when the temperature was up to 1450°C.The diffraction peaks of Al_2O_3 was still detected in the sample, as Fig.4b shown.

However, when the sample was obtained at 1500°C, only AlN and small amounts of C were present without any other detectable peaks (see Fig.4c). The presence of excess carbon in the reaction product possibly prevented the growth of AlN particle [23].

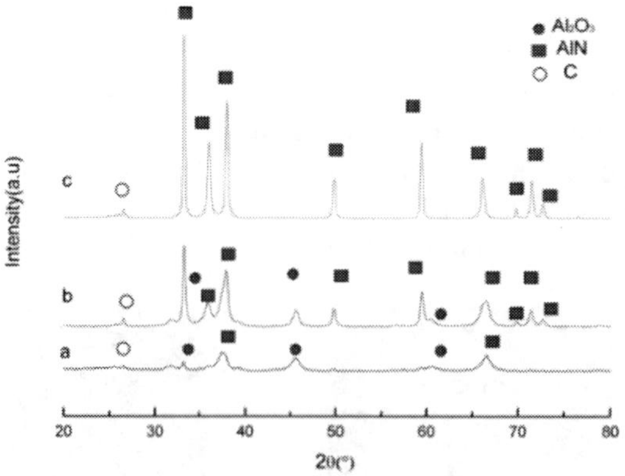

Figure 4. XRD patterns of reaction products obtained at different temperature for 2h.

3.4 The superfine AlN powder

AlN powder was prepared by carbon thermal reduction method sintering 1500°C for 2h using AlN precursor as raw material. The superfine AlN powder can be obtained by reaction product sintering at 650°C for 4h to remove excess carbon.Fig.5 shows the SEM image of AlN powder synthesized at 1500°C for 2h. No abnormal large particle can be observed and the average particle size was likely to be 40-80nm. Since the presence of excess carbon in the mixture possibly inhibited the growth of AlN grain [23], the particles of synthesized AlN powder observed in Fig.6 are very fine.

The XRD patterns of the final AlN powders are shown in Fig. 5, the main peaks can well correspond to the characteristic peaks of hexagonal AlN. It can be seen clearly that single AlN phase was found in the synthesis powder calcined at 1500°C heated for 2h. It can be seen clearly that single AlN phase was found in the synthesis powder calcined at 1500 heated for 2h. Compared with the sintering temperature of conventional synthesis method, the powder prepared by this process was purity and the sintering temperature was lower.

Figure 5. SEM image and XRD pattern of AlN powder synthesized at 1500°C for 2h

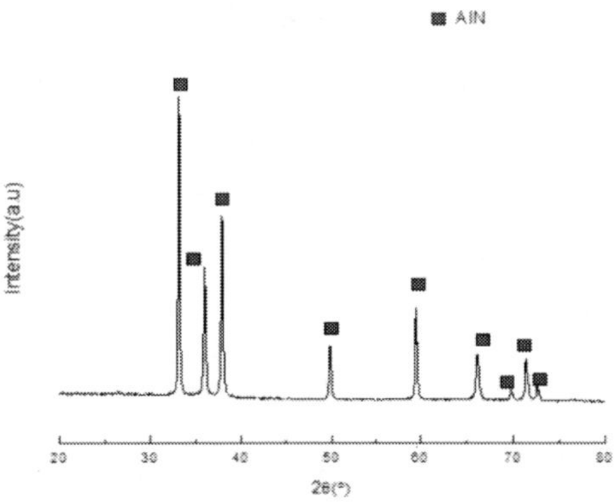

Figure 6. XRD pattern of AlN powder synthesized at 1500°C for 2h

4 Conclusions

In this paper, aluminum and carbon source were homogeneously distributed in the AlN precursor with sucrose. Optimum process for AlN powder which prepared by carbon thermal reduction method was 1500°C for 2h followed by heating at 650°C for 4h to remove any extra carbon. The superfine AlN powder with smaller size was successfully synthesized by using above process. Meanwhile, this synthesis process is superior to the traditional preparation process with single aluminum and carbon source. It can lower the reaction temperature and reduce heating time for conversion of AlN precursor. And the process in this paper could be suitable to the preparation of high density ceramics.

References

1. Hashimoto N, Sawada Y, Bando T, et al. Preparation of Aluminum Nitride Powder from Aluminum Polynuclear Complexes [J]. Journal of the American Ceramic Society, 1991, 74(6):1282-1286.
2. Tsuge A;Inoue H; Kasori M, et al. Raw material effect on AIN powder synthesis from Al2O3 carbothermal reduction. Journal of Materials Science. 1990, 25, 2359-2361.
3. Qin M L, Du X L, Li S, et al. Preparation of AlN powders from precipitated precursor [J]. Chinese Journal of Nonferrous Metals, 2005.
4. Dai C, Zhang X, Liu S. Study of Mechanism for Microwave Synthesis of Ultrafine AlN Powder [J]. Journal of Inorganic Materials, 1997, 12(5):755-758.

5. Komeya K; Mitsuhashi E; Meguro T. Synthesis of AlN Powder by Carbothermal Reduction-Nitridation Method: Effect of Additives on Reaction Rate [J]. Journal of the Ceramic Society of Japan. 1993, 101:377-382.
6. G. B. Schaffer; P. G. Mc Cormic. Reduction of oxides by mechanical alloying. Appl Phys Lett.1989, 55, 45-46.
7. Calka A; Nikolov J I;Wantenaar G H J. Low temperature synthesis of Al-AlN composites from a nanostructure made by controlled magneto-ball milling of Al in ammonia. Journal of Applied Physics, 1994, 75, 4953-4955.
8. Peng-Liang L I, Zhou J E, Sheng-Qi X I. Cubic AlN Synthesized by High Energy Ball Milling and Its Phase Conversion at High Temperature [J]. Journal of Inorganic Materials, 2006, 21(4):821-827.
9. Liu X, Ma M, Xi S. Study on mechanical activation synthesis of nano-crystal aluminium nitride [J]. Journal of the Chinese Ceramic Society, 2000.
10. Yan-Hong M A, Chen W. Preparation of AlN by the Sol-Gel Method [J]. Vacuum Electronics, 2009.
11. Xue li D U, Qin M L, Sun W, et al. Preparation and Sintering of Nanometer AlN Powders[J]. Vacuum Electronics, 2006.M.L.Qin; X.H.Qu; J.L.Lin. synthesis of ultrafine aluminium nitride powder by sol-gel process. Mater Sci Eng powder Metal.2002, 7, 51- 53.
12. Baraton M I, Chen X, Gonsalves K E. FT-IR surface characterization of a nanostructured aluminum nitride powder prepared via sol-gel chemical synthesis [J]. 1995.
13. Nur Fahana M A; Lee Z Y; Fong C Y, et al. Growth and Characterization of Aln Thin Film Deposited by Sol-Gel Spin Coating Techniques. Advanced Materials Research.2015, 1107:667-671.
14. Kimura I; Ichiya K;Ishii M, et al. Synthesis of fine AlN powder by a floating nitridation technique using an N2/NH3 gas mixture. Journal of Materials Science Letters.1989, 8,303-304.
15. Hwang C C, Lin C S, Wang G P, et al. A self-propagating high-temperature synthesis method for synthesis of zinc oxide powder. Journal of Alloys & Compounds, 2009, 467(1–2):514-523.
16. Y.M.Zhao; G.S.Sun; J.Y, .Lee,etal. Growth of a Novel Periodic Structure of SiC/AlN Multilayers by Low Pressure Chemical Vapour DepositionGrowth of a Novel Periodic Structure of SiC/AIN Multilayers by Low Pressure Chemical Vapour DepositionChinese Physics Letters.2007, 24, 1753-1755.
17. Dollet A, Casaux Y, Chaix G, et al. Chemical vapour deposition of polycrystalline AlN films from AlCl 3 –NH 3, mixtures. : Analysis and modelling of transport phenomena [J]. Thin Solid Films, 2002, 406(1-2):1-16.
18. Sun Y; Zhu R; Ye H, et al. Towards a bioinformatics analysis of anti-Alzheimer's herbal medicines from a target network perspective. Briefings in Bioinformatics, 2013, 14,327-343.
19. Cimalla V, Foerster C, Cengher D, et al. Growth of AlN nanowires by metal organic chemical vapour deposition. Physica Status Solidi, 2006, 243, 1476–1480.
20. Zhang D. Synthesis of AlN Powders by Carbonthermal Reduction Method [J]. Non-metallic Mines, 2001.
21. Wang J;Wang W L; Ding P D, et al. Synthesis of cubic aluminum nitride by carbothermal nitridation reaction. Diamond & Related Materials, 1999, 8, 1342-1344.
22. Pankratz, L. B., Stuve, J. M., Gokcen, N. A., & Mines., U. S. B. O. (1984). Thermodynamic data for mineral technology.
23. L.C. Pathak; A. K.Ray;S.Das; C.S.Sivaramakrishnan, P.Ramachandrarao,J.Am.Ceram.Soc. 1999, 82, 257-260.

Electro Polymerization of Polypyrrole Coatings Doped with Different Proton Acids for Corrosion Protection of 304 Stainless Steel

Li Jiang, Haibo Ma, Jiaqi Zhang, Yu Lu, Hongbin Lu[a] and Xiangkang Meng[a]

Institute of Materials Engineering, National Laboratory of Solid State Microstructures, College of Engineering and Applied Sciences, Nanjing University, Jiangsu, People's Republic of China

Abstract. Polypyrrole (PPY) coatings doped with sulphuric, oxalic and phosphoric acid were prepared on 304 stainless steel (304SS) by electropolymerization. Surface morphology and anti-corrosive performance of the polymer film was investigated by scanning electron microscope and electrochemical tests. The results indicate that the phosphoric acid doped PPY coating has a compact and smooth surface which shifts the E_{corr} of 304SS towards to more positive value, while the j_{corr} was dropped from 10.4 to 0.88 μA cm^{-2}. Long-term immersion test shows that the PPY coating doped with phosphoric acid provides better and durable corrosion resistance in comparison with the PPY coatings doped with the sulphuric and oxalic acids.

1 Introduction

Stainless steel (SS) becomes one of the most suitable metallic materials for industrial applications in the contemporary society due to the advantages of decent conductivity, high mechanical strength, low gas permeability and reasonable cost [1, 2]. However, the main problem associated with SS is to resist corrosion in the aggressive medium [2, 3]. The occurrence of corrosion not only brings large economic loss, the dissolution of SS also leads to harmful metallic ions (Cr, Fe, Ni et al.) release out, which undoubtedly contributes to the environmental pollution. Therefore, surface modification coatings have been proposed to improve the corrosion resistance of SS [4, 5]. Since Deberry [6] found conducting polyaniline coating could provide corrosion protection for steel, different conducting polymers for example polyaniline (PANI) [7, 8] and polypyrrole (PPY) [9, 10] are intensively investigated for their corrosion protection applications[11, 12]. It is widely accepted that the polymer helps passivate the surface between the coating and the SS substrate[13], moreover, the conducting polymer can be regarded as barrier layer to protect the SS substrate from corrosive ions. PPY is one of the most studied conducting polymers due to its good anti-corrosive performance and easily preparation nature [14]. Commonly the conductivity of PPY can be achieved by doping protonic acid during the electro polymerization process and usually the doped acid has significant effect on conductivity, capacitive characteristic and corrosion resistance of PPY [9].

[a] Corresponding author: luhb@nju.edu.cn/mengxk@nju.edu.cn

In order to get a homogenous, compact PPY coating for the use 304 SS to decrease the economic loss caused by corrosion, in this study PPY was electro polymerized in sulphuric, oxalic and phosphoric acid solution respectively. Scanning electron microscope(SEM), electrochemical impedance spectroscopy, open-circuit potential and potentiodynamic polarization were used to explore the effect of different proton acid of the anti-corrosive properties for PPY coating in acid environment, and the protection mechanism of the PPY coating was discussed based on the results.

2 Experimental procedures

The 304SS plate used as work electrode was sealed with epoxy resin with an exposed surface area for 1 cm^2 and polished with emery papers, followed by rinsed with deionized water and acetone, respectively. The composition of type 304SS is shown in Table 1 [15]. Three electrode system was carried out for electro polymerization process, with saturated calomel electrode (SCE) as reference electrode and a platinum plate as counter electrode. The aqueous solutions for electro synthesis consist of 0.1 M pyrrole and 0.3 M proton acid. Using cyclic voltammetry (CV) technique to achieve the electro polymerization of PPY films, the potential varied from -700mV to 1800mV versus SCE with potential scan rate of 10 mV^{-1} for 12 cycles. The anti-corrosion properties of the PPY coatings were investigated in 0.3M H_2SO_4 solution through electrochemical test and SEM was used to observe the morphologies of the polymer coatings.

Table 1. The composition of type 304SS substrate

C	S	P	Si	Mo	Ni	Mn	Cr	Fe
0.05	0.028	0.015	0.59	0.30	8.12	0.88	18.5	Balance

3 Results and discussions

3.1 Electro synthesis of PPY coatings

Black PPY coatings electropolymerized with different proton acid are observed after CV process and Figure 1 shows their surface morphologies. All coatings consist of globular particles, which is in agreement to the previously reported PPY films [16]. It can be seen that PPY-H_2SO_4 film exhibits some tiny cracks while $H_2C_2O_4$ doped PPY coating shows relatively dense but not smooth surface as displayed in Figure 1a and Figure 1b.

However, the film of PPY-H_3PO_4 is observed the densest, continuous and crack free with a more homogenous surface morphology(Figure 1c), which may becuse pyrrole monomers has better solubility in phosphoric acid solution. The smooth and crack free surface of PPY- H_3PO_4 indicating that this coating has the finest grain size, thus can offer the best physical barrier to 304SS substrate to isolate the corrosive ions.

Figure 1. The morphologies of PPY coatings doped with H_2SO_4 (a), $H_2C_2O_4$ (b) and H_3PO_4(c)

3.2 Potentiodynamic polarization measurements

Figure 2a shows the polarization curves for uncoted and coated 304SS after 1 h immersion in 0.3M H_2SO_4 solution. Using Tafel extrapolation method to calculate the samples corrosion potential (E_{corr}) and corrosion current (j_{corr}). The potentiodynamic polarization parameters are partly summarized in Table 2, the lower j_{corr} values demonstrate the decrease in corrosion rate and the more positive E_{corr} values attributes to the substrate which is difficult to corrode in the aggressive environment[17]. The E_{corr} and j_{corr} values for bare 304SS are -315 mV (SCE) and 10.4 μA cm^{-2} respectively, it is observed that all the coatings have lower j_{corr} with more positive E_{corr} values compare with the 304SS substrate due to their anti-corrosive nature. Especially H_3PO_4 doped PPY coating, which has the most positive E_{corr} and smallest j_{corr} values of 58 mV and 0.88 μAcm^{-2}, respectively, suggesting that PPY-H_3PO_4 film exhibits outstanding corrosion protection performance. Besides, the fitting polarization resistance R_p for the uncoated and coated 304SS based on the polarization data in 0.3 M H_2SO_4 solution can be acquired through the following equation :

$$R_p = \frac{\beta_a \beta_c}{2.3 j_{corr} (\beta_a + \beta_c)} \quad (1)$$

where β_a and β_c represants the Tafel slopes for the anodic and cathodic reactions, respectively. It can be see that R_p of coated subtrate all considerably higher than that of bare 304SS, in particular, the R_p value of the PPY-H_3PO_4 film is the largest among all samples. The results explicitly shows that H_3PO_4 doped PPY coating can offer better protectiveness than PPY-H_2SO_4 and PPY-$H_2C_2O_4$ coatings, which corresponding to the densest morphology shown in Figure 1.

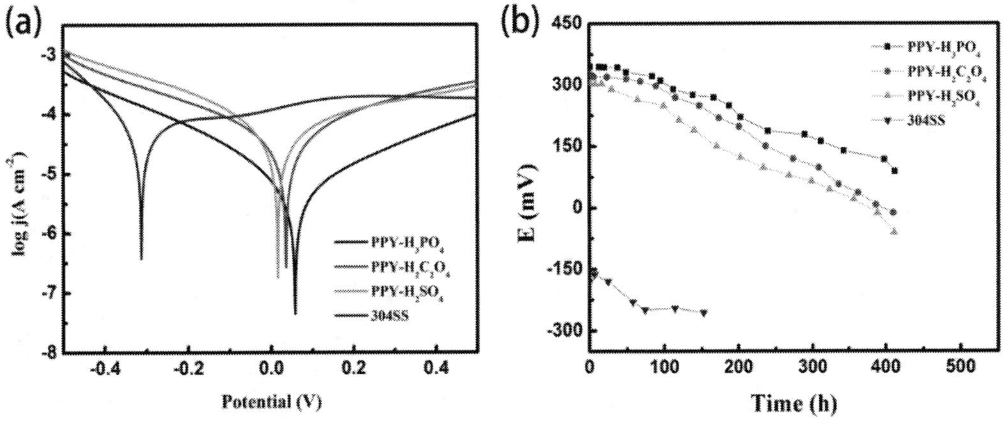

Figure 2. Tafel polarization curves of bare and different PPY coated 304SS (a) and time dependence of E_{ocp} for different samplesin 0.3 M H_2SO_4 solution during long-term immersion (b).

Table 2. Polarization parameters of the bare and different PPY coated 304SS in 0.3 M H_2SO_4 solution

Samples	β_a	β_c	E_{corr}(mV)	i_{corr}(A cm^{-2})	R_p(Ω cm^2)
304SS	0.042	0.063	-315	1.04E-5	1065
PPY-H_2SO_4	0.067	0.071	16	8.47E-6	1763
PPY-$H_2C_2O_4$	0.049	0.043	49	2.33E-6	4270
PPY-H_3PO_4	0.045	0.058	58	8.81E-7	12421

DOI: 10.1051/matecconf/201710903007

3.3 Open-circuit potential (E_{ocp}) measurements

The curves of E_{ocp} versus time of uncoated and different PPY coated 304SS are presented in Figure 2b. The increased E_{ocp} values for the bare 304SS in the initial immersion stage attributes to the formation of passive oxide layer and then the value decreased to -250mV, indicating that the passive layer gradually destroyed and the substrate suffered from corrosion[11]. The initial E_{ocp} values of all PPY coatings are more than 300mV (SCE) higher than the value of uncoated 304SS substrate. Initially, the E_{ocp} values for PPY coatings relatively stable, however the E_{ocp} values for PPY-H_2SO_4 and PPY-$H_2C_2O_4$ coatings significantly decreased with the prolonged immersion time, which attributes to the slow degradation of these two coatings and electrolyte gradually penetrate inward. The E_{ocp} values for PPY-H_3PO_4 coating fall slightly with time and always larger than other PPY coatings, suggesting that PPY-H_3PO_4 coating can prevent corrosive ions from penetration effectively during the long-term immersion and provides better corrosion protect for 304SS as compared to PPY-H_2SO_4 and PPY-$H_2C_2O_4$ coatings.

3.4 Electrochemical impedance measurements

Using the electrochemical impedance spectroscopy (EIS) measurements to obtain better assessment of the corrosion protection properties for the coatings. Nyquist plots for the three kinds of coatings protected 304SS after 1 day and 16 days immersion are presented in Figure 3.

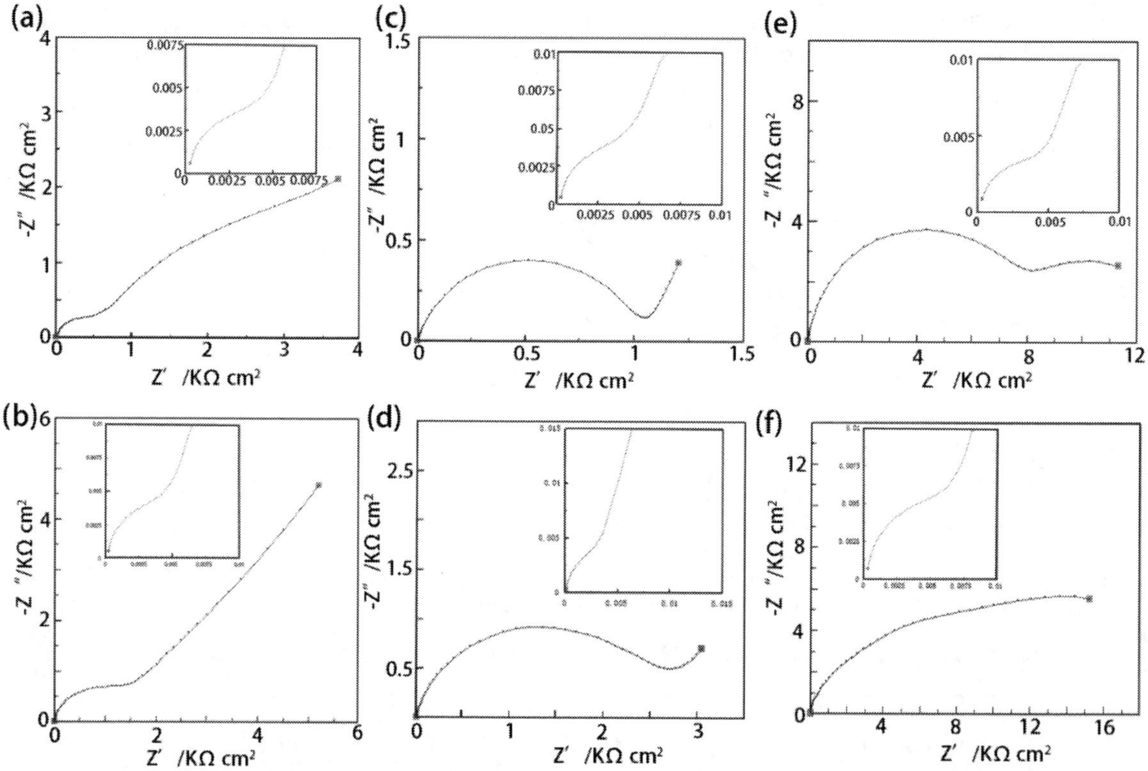

Figure 3. The Nyquist plots of 304SS coated by PPY-H_2SO_4 (a), (b); PPY-$H_2C_2O_4$ (c), (d) and PPY-H_3PO_4 (e), (f) in 0.3 M H_2SO_4 solution after 1 day and 16 days immersion, respectively.

The plots of PPY-H_2SO_4 and PPY-$H_2C_2O_4$ coated 304SS all consist of two depressed capacitive loops in high frequency region with a linear part in low frequency region (Figure 3a -3d), which shows the diffusion of electrolyte. Generally, the high frequency parts reflects the coatings information and the low-frequency parts tells the processes at the coating/304SS interface in EIS plots[11]. Therefore, the lines from low frequency region of PPY-H_2SO_4 and PPY-$H_2C_2O_4$ coatings closed to 45 degree after 16 days immersion(Figure 3b, 3d) shows that the surface of the coating destroyed to a certain extent that the corrosive ions have penetrate inward, which in correspondence with OCP data(Figure 2b).

However, the Nyquist curves of PPY-H_3PO_4 coated 304SS shows capacitive characteristic instead of diffusion process in the low frequency region during immersion (Figure 3e, 3f), and the capacitance loop in high frequency region is higher than other coatings with continuously increase through the corrosion process, indicating that PPY-H_3PO_4 exhibits preferable anti-corrosive performance among the three coatings.

In general, the conductive polymer could provides corrosion protection not only for its barrier effect, but also for anodic protection through passivate the substrate. However, when the doped anion released during the protection process, the polymer will be reduced slowly and just serve as a mechanical barrier in its reduction form. It is reproted that the electrodeposited polyaniline film in phosphoric acid solution can improve the adhesiveness of the coating on the metal surface[15], we can deduce that the electropolymerized PPY coating in phosphoric acid solution has similar properties. Therefore, the good adhesion performance and compact surface of PPY-H_3PO_4 coating offers an excellent physical barrier to isolate corrosive ions from the electrolyte solution. Meanwhile, the dedoping process may carried out difficultly due to the strong binding force between phosphate radical thus corrosion protection of the PPY-H_3PO_4 coating is more effective as a physical barrier and an stable oxidizer than PPY-H_2SO_4 and PPY-$H_2C_2O_4$ coatings.

4 Conclusions

The corrosion protection performances of the PPY doped with H_2SO_4, $H_2C_2O_4$ and H_3PO_4 were investigated in acid corrosive environment. The electrochemical tests showed that all the different coatings provide a certain corrosion protection for 304SS substrate. PPY coatings enlarged the corrosion potential of bare 304SS by more than 300 mV (SCE), particularly PPY-H_3PO_4 shifts the corrosion potential to a more positive value and significantly decreased the corrosion current density. Moreover, PPY doped with H_3PO_4 has a higher corrosion impedance compared to other coatings and the impedance increased continuously during the 16 days H_2SO_4 immersion, suggesting that the PPY-H_3PO_4 coating provides more effective corrosion protection for 304SS resulted from a physical barrier and an stable oxidizer.

Acknowledgements

This work was jointly supported by the National Natural Science Foundation of China (No. 51501088), the Fundamental Research Funds for the Central Universities (No. 021314380019), the Applied and Basic Research Project of Nantong (No. GY12015003), and the Science & Technology Support Plan of Jiangsu Province (BE2014039).

References

1. Y. Yu, S. Shironita, K. Nakatsuyama, K. Souma, M. Umeda, Appl. Surf. Sci. **388**, Part A (2016) 234-238.
2. C. Shanmugham, N. Rajendran, Prog. Org. Coat. **89** (2015) 42-49.
3. M. Mehdipour, R. Naderi, B.P. Markhali, Prog. Org. Coat. **77** (2014) 1761-1767.
4. M. Aparicio, A. Jitianu, G. Rodriguez, A. Degnah, K. Al-Marzoki, J. Mosa, L.C. Klein, Electrochim. Acta **202** (2016) 325-332.
5. J.-B. Cambon, F. Ansart, J.-P. Bonino, V. Turq, Prog. Org. Coat. **75** (2012) 486-493.
6. D.W. DeBerry, J. Electrochem. Soc. **132** (1985) 1022-1026.
7. J.A. Syed, H. Lu, S. Tang, X. Meng, Appl. Surf. Sci. **325** (2015) 160-169.
8. J.A. Syed, S. Tang, H. Lu, X. Meng, Colloids Surf., A **476** (2015) 48-56.

9. M. Gholami, P.M. Nia, Y. Alias, Appl. Surf. Sci. **357**, Part A (2015) 806-813.
10. T. Zhang, C.L. Zeng, Electrochim. Acta **50** (2005) 4721-4727.
11. Y.J. Ren, C.L. Zeng, J. Power Sources **182** (2008) 524-530.
12. T.J. Pan, X.W. Zuo, T. Wang, J. Hu, Z.D. Chen, Y.J. Ren, J. Power Sources **302** (2016) 180-188.
13. M.B. González, S.B. Saidman, Prog. Org. Coat. **75** (2012) 178-183.
14. R. Vera, R. Schrebler, P. Grez, H. Romero, Prog. Org. Coat. **77** (2014) 853-858.
15. Y. Gao, J.A. Syed, H. Lu, X. Meng, Appl. Surf. Sci. **360**, Part A (2016) 389-397.
16. D. Kowalski, M. Ueda, T. Ohtsuka, Corros. Sci. **49** (2007) 1635-1644.
17. B. Rikhari, S.P. Mani, N. Rajendran, RSC Advances **6** (2016) 80275-80285.

Design and Research of New Syntactic Foam Pressure Parts

Fang Xiong[1, a], Shen Xisheng[1], Chen Zhixing[1], Xu ZhiQiang[1], Jin JiaoHui[1]

[1]*Fishery Machinery and Instrument Research Institute, Chinese Academy of Fishery Sciences, Shanghai, China*

Abstract. With the development of materials and technology, new composite materials which can meet the requirement of new technology are produced, especially in some harsh conditions where traditional materials cannot meet the needs. In this paper, a new type of syntactic foam pressure part is studied, simulate and calculate the impact of load stress and breaking type by finite element analyze. The results of experiment are same as the calculation. A method is established in this paper to provide a reference and research ideas for the similar products.

1 Introduction

In order to meet the harsh conditions requirement and lower the cost, new syntactic foam material is applied to design and research of structural pressure parts, which gives full play to the characteristics of light weight, heat insulation and adjustable strength. Pressure parts are usually sealed together with the large structural parts, but with the impact of instantaneous load, if the preset pressure is reached, it is immediately break to remind the engineer of load conditions. It can protect the main structural parts. But so far, this kind of research is rare in China [1].

Previous research on the behaviour of syntactic foam mainly focused on the experimental research. On the one hand these researches meet the requirement of application at that time, and on the other hand, they also promote the improvement of the material preparation process. Not only the macroscopic mechanical behaviour of these new types of foams has been studied, but also the research on the microscopic failure mechanism of these materials has been paid more attention. However, the mechanical properties of the syntactic foams are closely related to the technological conditions and process level. Also the environment where the syntactic foams used is complex, so the study of this kind of new material is far from the end. Compared with the experimental research, the theoretical research work is obviously lacking. People can only use some of the existing research results in composite materials mechanics to give the estimated value of the material parameters or put forward some empirical formula.

This paper is based on the new syntactic foam products; the performance of this material is tested and the simulation software Ls-dyna is used to simulate the breaking type of the product under transient impact load. The results are verified by comparison with the experiment.

2 Material analyse

[a] Corresponding author: 13472617125@163.com

2.1 Material introduction

At the end of the 1960s, research on syntactic foams began. DeRuntz systematically studied the mechanical behavior of syntactic foam, He experimented on the hollow cylindrical specimen and the cylindrical specimen made of hollow ball-resin syntactic foam. The experiment was carried out by pulling, pressing and twisting the uniaxial experiment, biaxial stress experiment, and triaxial stress experiment under water pressure.

In the mid-1970s, Barber and others firstly carried out the three-phase syntactic foam mechanical behavior research work, they put hollow glass microspheres in the amino rigid foam. It is obviously improved the compression and bending strength and modulus, also improved the impact resistance and the dimensional stability. Due to the addition of glass microspheres, the lower density of the syntactic foam can achieve higher density. So it provide possibility to use lighter syntactic foam in structural applications.

In the early 1980s, M. Narkis and others systematically studied the preparation of two-phase and three-phase syntactic foams and their corresponding mechanical properties. They used two different types of microspheres (phenolic and glass microspheres) to fill the foam and it showed good compression mechanics. In addition, M. Narkis and others also studied the rotational molding of thermosetting three-phase syntactic foams and the compressive strength and modulus energy of the flexible syntactic foam are given, also the temperature effect was studied for polyimide based foams and the concept of stress transfer efficiency is proposed. The Ishai and Cohen equations and the Lewis and Nielsen equations were modified by the stress transfer efficiency parameters. In order to further study the relationship between mechanical properties and density, they also studied the tensile and bending behavior of the three-phase silicone syntactic foam. The results showed that the tensile strength of the foam is lower than its bending strength and the bending strength is much lower than the compressive strength. Under the same matrix volume content, the hollow ball composite foam has a higher compressive strength but the elastic modulus of the solid spherical foam is much higher than that of the foam containing hollow spheres.

In this paper, pressure parts are made of syntactic foam. It is a new kind of structural and functional composite material which is filled with hollow microspheres and foaming agent in the resin matrix. Compared with traditional foam, the syntactic foam has a high specific strength and rigidity at low density due to the enhancement of hollow microspheres; it also has better thermal stability, insulation, corrosion resistance and etc. [2][3]

2.2 Material performance test

In order to describe the dynamic breaking behavior of the pressure parts accurately by finite element simulation, the key mechanical parameters of the material must be obtained by experiments. The syntactic foam is subjected to a quasi-static uniaxial tensile test at room temperature to measure a number of valid data. The loading strain rate is kept constant by 3mm/min and the corresponding strain rate is 0.001/s. To measure the elastic modulus and Poisson's ratio of the material, both strain values of vertical and horizontal sides should be recorded when measuring the force on the specimen. Non-contact measurement is used to avoid the interference of elastic factors caused by mechanical clamp and strain gauges, it can ensure the measurement accuracy [4][5]. As shown in Figure 1.

Figure 1. Tensile test and strain measurement experiment

DOI: 10.1051/matecconf/201710903008

2.3 Analysis of test results

There is no necking phenomenon during the process of stretching. Most of them are suddenly broken which are perpendicular to the direction of stretching. The section is flat and the material exhibits brittleness. The material exhibits linear elasticity at the initial stage of the bearing, and gradually becomes non-linear as the load increases. What the syntactic foam exhibits is following the theory of continuous elasticity in solid mechanics. After averaged the effective data of the sample, the results are shown as below: elasticity modulus is 758MPa, the Poisson's ratio is 0.329, the fracture stress is 8.05MPa, and the fracture strain is 1.22%.

3 Simulations

3.1 Pressure part model

Draw a 3D model of pressure part in Solidworks and divide it, as shown in Figure 2 and 3:

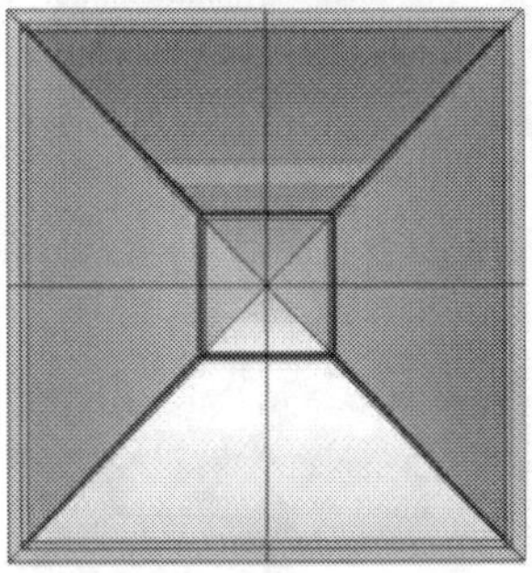

Figure 2. Bottom view

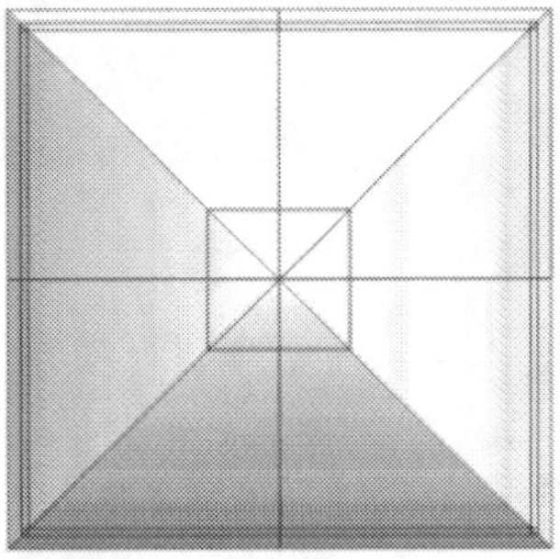

Figure 3. Top view

3.2 Unit selection and meshing

Select the solid 3D solid164 module and do the hexahedral and tetrahedral meshes. As shown in figure 4 and 5.

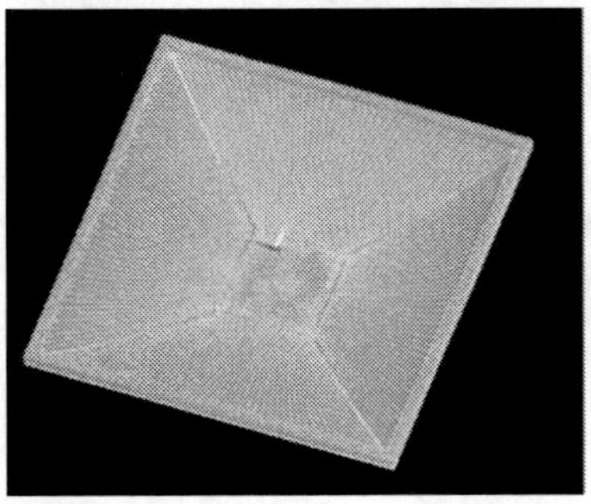

Figure 4. Overall grid

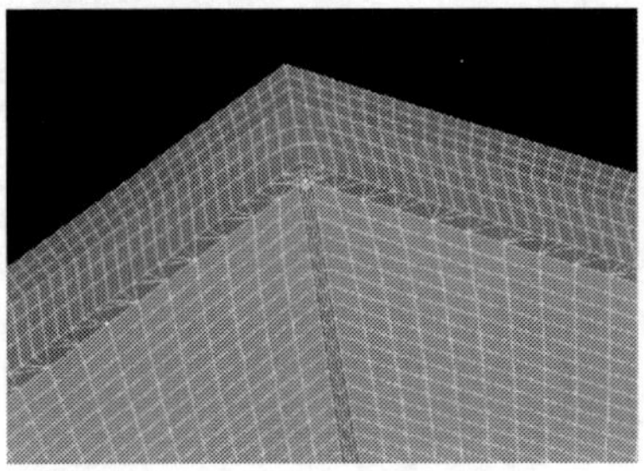

Figure 5. Local grid

3.3 Boundary conditions

According to the real situation, constrain the pressure part.

3.4 Apply the load

Apply impact load on the pressure part and the instantaneous impact load is P=0.075MPa.

3.5 Calculation

The pressure part is subjected to an instantaneous impact load and the broken position is the location of the fracture groove and the part is completely broken into 5 blocks. As shown in figure 6 and 7.

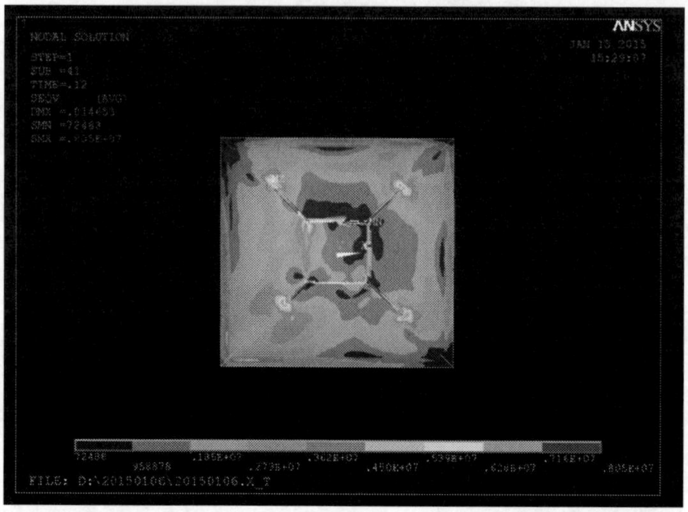

Figure 6. The stress distribution of the moment of cracking

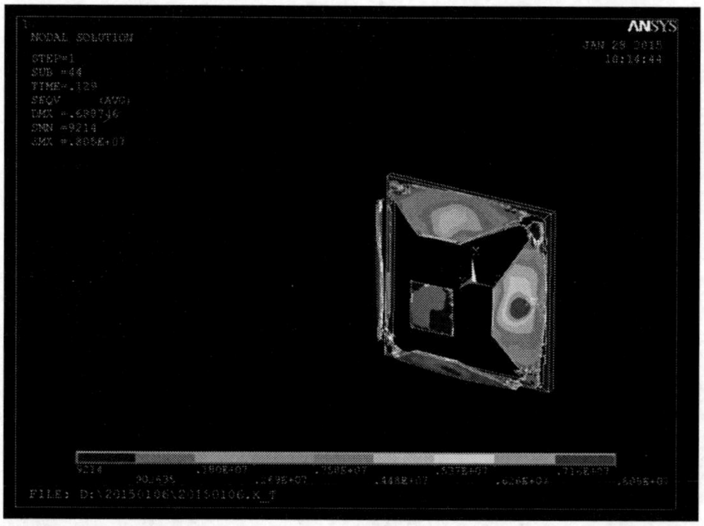

Figure 7. Debris distribution of the pressure part

4 Impact breaking test

In order to verify the validity of the simulation, an impact breaking test is carried out. The pressure part is fixed on the left and there is a gas tank with high pressure gas on the right. To compare results with the simulation, there are three pressure sensors in the chamber, to monitor pressure change during the test process. When the test begins, valve will be opened quickly, then, high pressure gas released. Under the action of the impact load, the pressure part is quickly broken and the whole process of the experiment is finish.

To ensure the accuracy of the experiment, the test was repeated several times. Four times' experimental data are shown in table 1.

DOI: 10.1051/matecconf/201710903008

Table 1. Test data

No.	Breaking strength（MPa）	Fractures and pieces	Discrete rate
1	0.072	Break along the trench, 5 pieces	-4.16%
2	0.077	Break along the trench, 5 pieces	2.59%
3	0.067	Break along the trench, 5 pieces	-11.94%
4	0.07	Break along the trench, 5 pieces	-7.14%

5 Calculation and results analyze

In this experiment, a total of four impact breaking tests were carried out, the breaking forms of all the pressure parts were broken along the trench and broken into 5 pieces. Compared the results with the simulation, the theoretical calculation and experimental results are almost same in broken form, fragment size, and fragment shape.

6 Conclusions

Simulate the transient impact of the syntactic foam pressure parts by finite element simulation software, insert the data to Ls-dyna and calculate the broken form, fragment size and fragment shape Then, complete the impact breaking test and compare the results with the simulation and the results show they are almost same. The calculation can be drawn that this new syntactic foam material can meet the requirement of design and working conditions well, also the producing process of the material is simple and the cost is low. The simulation method and results are established in this paper to provide a reference and research ideas for the syntactic foam pressure parts.

References

1. Y.H. Wu, X.H. Wang, B.J. Ma Structure & Environment Engineering, **34,** No.6, 10-12(2007)
2. Z.X. Lu, S. Wang, Z.M. Li, Chinese Journal of Aeronautics, No.5, 800-802(2006)
3. Z.X. Lu, Y.L.Yuan, Acta Materiae Compositae Sinica, **19,** No.5, 115-116(2002)
4. Z.X. Lu, C.H. Kou, H.X. Li, Journal of Beijing University of Aeronautics and Astronautics, **24**, No.6, 646-649(1998)
5. M. Yu, P. Zhu, Y.Q. Ma, Acta Materiae Compositae Sinica, No.3, 226-228(2013)

Piezoelectric Response Evaluation of ZnO Thin Film Prepared by RF Magnetron Sputtering

Da-Long Cheng[1], Kuo-Sheng Kao[1], Chia-Hua Liang[2], Yu-Che Wang[1], Yu-Chieh Chen[1], Wei-Che Shih[3] and Leong-Perng Chan[4, 5,a]

[1]*Department of Computer and Communication, Shu-Te University, Taiwan*
[2]*Department of Cosmetic Science, Chia Nan University of Pharmacy and Science, Taiwan*
[3]*Department of Electrical Engineering, National Sun Yat-Sen University, Kaohsiung, Taiwan*
[4]*Institute of Clinical Medicine, Kaohsiung Medical University, Kaohsiung, Taiwan*
[5]*Department of Otolaryngology-Head and Neck Surgery, Kaohsiung Medical University Hospital, Kaohsiung Medical University, Kaohsiung, Taiwan*

Abstract. The most important parameter of piezoelectric materials is piezoelectric coefficient (d33). In this study, the piezoelectric ZnO thin films were deposited on the SiNx/Si substrate. The 4 inches substrate is diced into 8 cm× 8 cm piece. During the deposition process, a zinc target (99.999 wt%) of 2 inches diameter was used. The vertical distance between the target and the substrate holder was fixed at 5 cm. The piezoelectric response of zinc oxide (ZnO) thin films were obtained by using a direct measurement system. The system adopts a mini impact tip to generate an impulsive force and read out the piezoelectric signals immediately. Experimentally, a servo motor is used to produce a fixed quantity of force, for giving an impact against to the piezoelectric film. The ZnO thin films were deposited using the reactive radio frequency (RF) magnetron sputtering method. The electric charges should be generated because of the material's extrusion. This phenomenon was investigated through the oscilloscope by one shot trigger. It was apparent that all ZnO films exhibit piezoelectric responses evaluated by our measurement system, however, its exhibit a significant discrepancy. The piezoelectric responses of ZnO thin film at various deposition positions were measured and the crystal structures of the sputtering pressure were also discussed. The crystalline characteristics of ZnO thin films are investigated through the XRD and SEM. The results show the ZnO thin film exhibits good crystalline pattern and surface morphology with controlled sputtering condition. The ZnO thin films sputtered using 2 inches target present various piezoelectric responses. With the exactly related position, a best piezoelectric response of ZnO thin film can be achieved.

1 Introduction

Piezoelectric materials have been attracting considerable attention because of their potential applications for acoustic devices [1], micro motors[2], AFM cantilevers [3], UV detectors [4-5] or chemical sensors [6]. The most important parameter of piezoelectric materials is piezoelectric coefficient (d33). Conventionally, the techniques for piezoelectric coefficient measurement can be

[a] Corresponding author: oleon24@yahoo.com.tw

divided into two categories: direct measurement and indirect measurement.In this study, we takes focus on the direct measurement method of piezoelectric coefficient by using a homemade stable force generator. Firstly, a servo motor is used to produce a fixed quantity of force, for giving animpact against to the piezoelectric film. The electric charges should be generated because of the material's extrusion. This phenomenon will be investigated through the oscilloscope by one shottrigger. The influences caused from various film thickness have been studied. The responses of the piezoelectric film are expected to be monitored under the condition with quantitative force. Finally, a reliable platform of piezoelectric coefficient measurement will be established by suitable mechanical modificationdesign.

2 Experimental

2.1 Thin Film process

In this study, the piezoelectric ZnO thin films were deposited on the SiNx/Si substrate. The 4 inches substrate is diced into 8 cm× 8 cm piece. During the deposition process, a zinc target (99.999 wt%) of 2 inches diameter was used. The vertical distance between the target and the substrate holder was fixed at 5 cm. In order to illustrate the sample position, we defined the value of R/r in the sputtering chamber, where r is the radius of zinc target and R is the distance between the sample and the substrate holder centre, as shown in Fig.1. By varying the value of R/r, the influence of the geometry inside the chamber on the thin film properties can be realized.

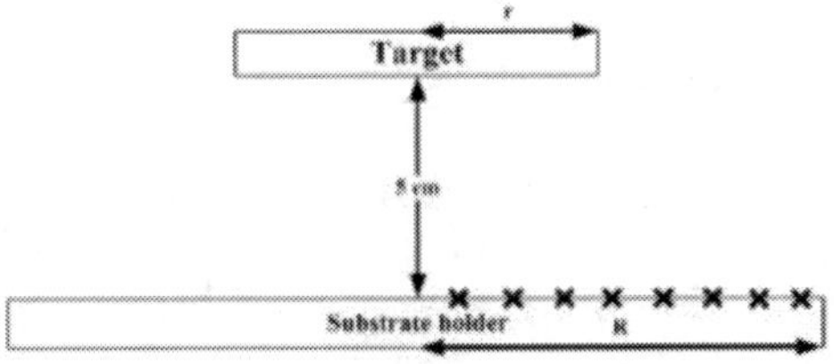

Figure 1. Schematic illustration of the sample position in the sputtering chamber.

Deposition parameters of various conditions were show in Table 1. ZnO thin films were deposited using the reactive radio frequency (RF) magnetron sputtering method. In order to obtain the high c-axis orientation ZnO thin films characteristics, two-step ZnO thin films deposition method was adopted to improve ZnO thin films characteristics. The ZnO thin films characteristics were determined by X-ray diffraction (XRD) using a Bruker D8 Advance with CuKα radiation. Cross-sectional morphologies of the ZnO thin films were observed by a scanning electron microscopy (JEOL6700 FESEM).

Table 1. Deposition parameters

Targets	Zinc (99.999 wt. %) 2 inch
Substrate	SiNx/Si
Base pressure (Torr)	$< 1\times10^{-6}$
Working pressure (mTorr)	10, 15, 25
Sputtering power (W)	85
Deposition time (min)	60
Substrate temperature (°C)	R. T.
$O_2/(O_2+Ar)$ (%)	75

2.2 Measurement Instrument

Figure 2 shows the experimental setup to fabricate the direct measurement system of piezoelectric response. The setup consists of a mini impact probe, a cantilever, a servo motor, an oscilloscope, a translation stage, and specimens to be tested. An impact probe driving by a servo motor connected with a stainless steel rod is submitted close to the sample surface and in contact with substrate. The probe with a metal tip acts perpendicularly to the thin film's surface to generate electrical pulse. In turn, due to the piezoelectric effect, the mechanical displacement of the film produces output electrical signal at the tip. In the meantime, the direct piezoelectric voltage can be measured by the impact probe.

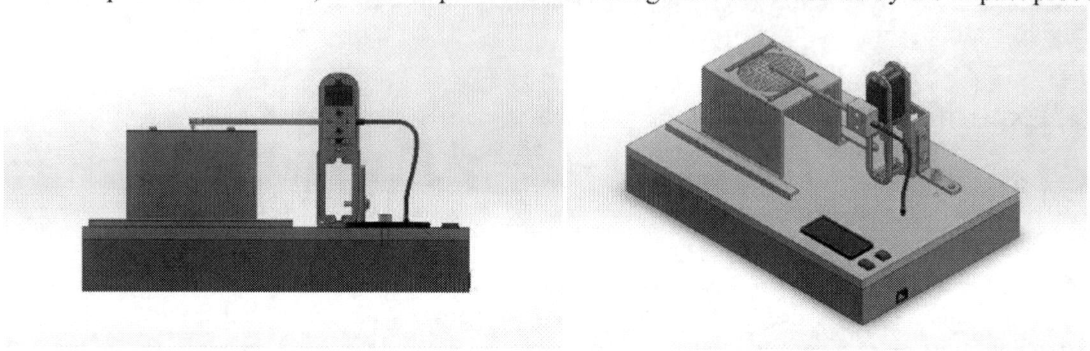

Figure 2. Experimental setup.

3 Results

3.1 Influence of the sputtering positions

As is shown in Fig. 3, a concentric ring pattern of rainbow colours forms because the different wavelengths of light interference at different film thicknesses. An obviously Newton's rings presented that a clear and definite discrepancy of thin film thickness is without a doubt. It can be concluded that the thickness of the ZnO film is decreased as the sample is placed far away from the target. In other words, the insufficient supplement of the ions results in the lack of atoms for the ZnO growth, leading to a thinner film thickness.

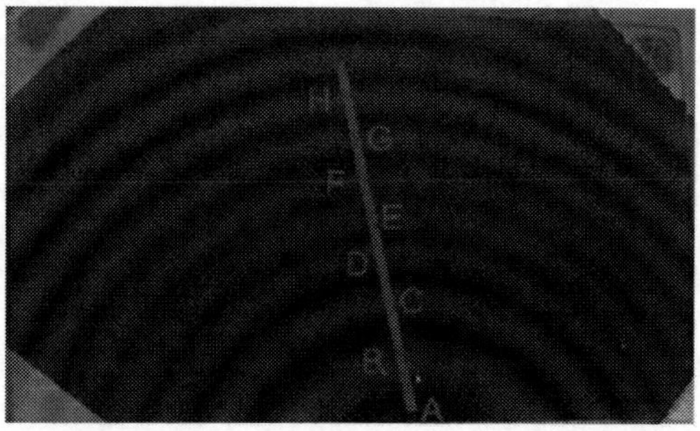

Figure 3. The top-view of a half substrate

3.2 Influence crystal structure of the sputtering pressure

Firstly, the sputtering pressure was varied to determine the optimal growth parameters of the ZnO thin films. The sputtering parameters were shown in table 1, the substrate temperature was set room temperature and sputtering pressure was set to 10 mTorr, 15 mTorr and 20 mTorr to investigate ZnO thin films of crystal structure. Fig. 4 shows XRD pattern of ZnO thin films deposition on the SiNx/Si substrate, revealing the excellent ZnO (002) orientated crystals around 34.5°. Even though XRD of the sputtering pressure at 15 mTorr was presented the highly c-axis oriented ZnO (002), the SEM of Fig. 5(b) shows good surface structure. Figure 5 exhibited ZnO thin films surface morphologies with various sputtering pressure, the Fig. 5(a) and (c) show disorder crystal structure. Besides, the Fig. 5(b) presented the uniform cobblestone crystal and identical grain. It exhibited the thin film has higher the ZnO (002) diffraction peak, the strongly c-axis (002) orientation of a piezoelectric thin films that has the higher electromechanical coupling coefficient [7].

As a result, the optimal sputtering pressure of 15mTorr, RF power of 85 W and O_2/O_2+Ar ratio of 75% were used to deposit ZnO thin films. The study result shows a strong c-axis-orientated crystallization, uniform grain size, well-textured columnar structures of the ZnO thin film.

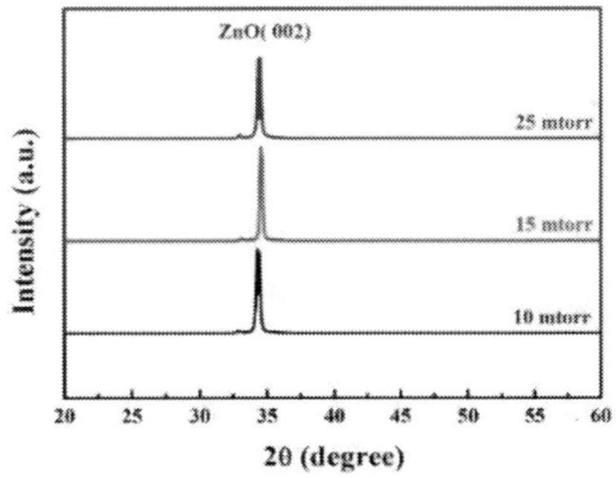

Figure 4. The XRD pattern with various sputtering pressure of (a)10 mTorr (b)15 mTorr (c)20 mTorr

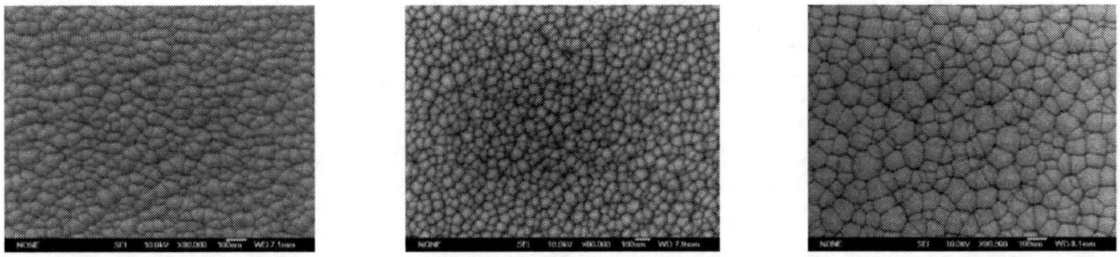

(a) (b) (c)

Figure 5. The surface morphologies with various sputtering pressure of (a) 10 mTorr (b) 15 mTorr (c) 20 mTorr

3.3 Piezoelectric response with related position

Figure 6 shows the piezoelectric responses of these ZnO films at various positions. The experimental data are directly captured from the direct measurement system. It is apparent that all ZnO films exhibit piezoelectric responses evaluated by our measurement system, however, its exhibit a significant discrepancy. Figure 7 shows the variations in the peak voltages of the responses as a function of sample position. From our observation, the increase in the peak voltage of ZnO films can be mainly attributed to the increment in the film's quality.

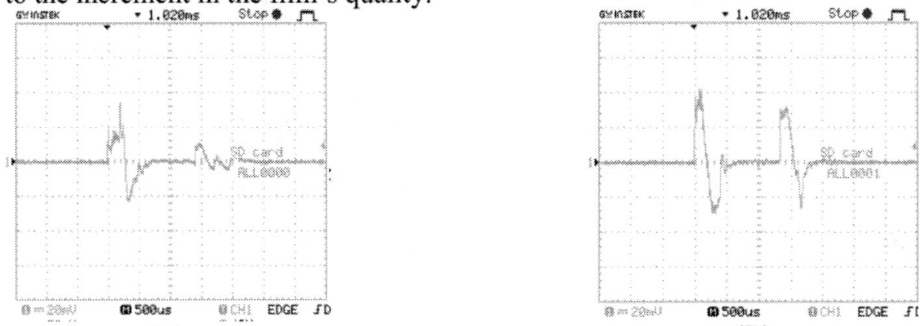

(A) (B)

DOI: 10.1051/matecconf/201710904001

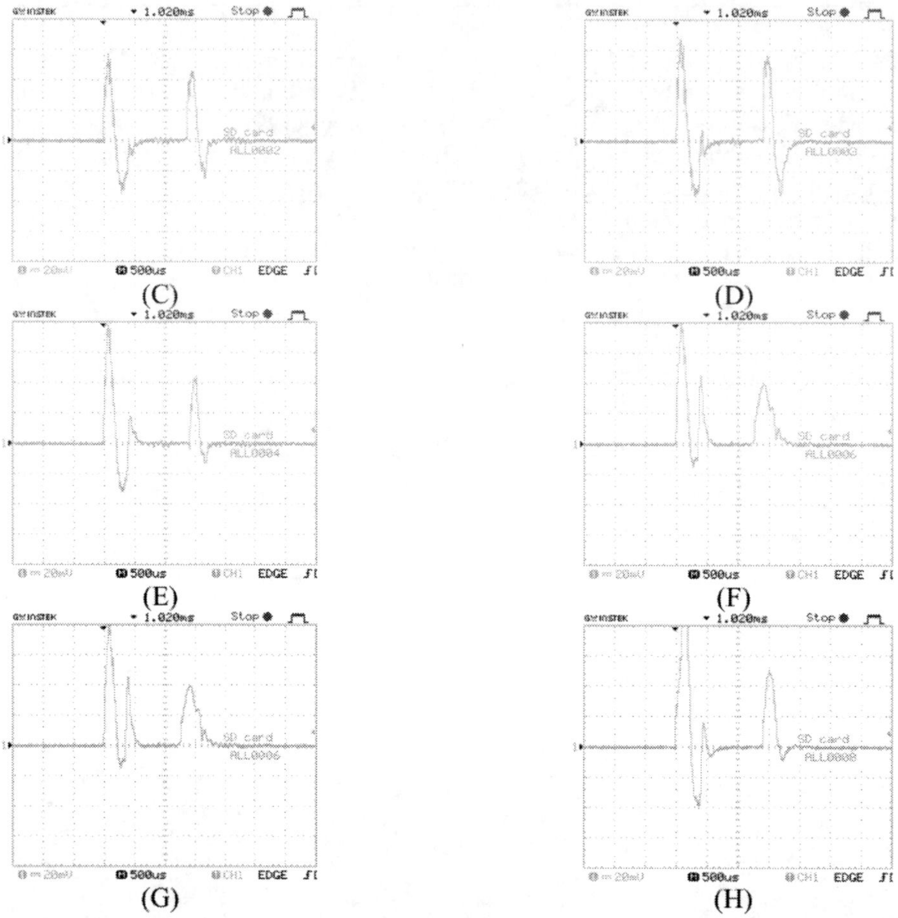

Figure 6. Piezoelectric response at various positions

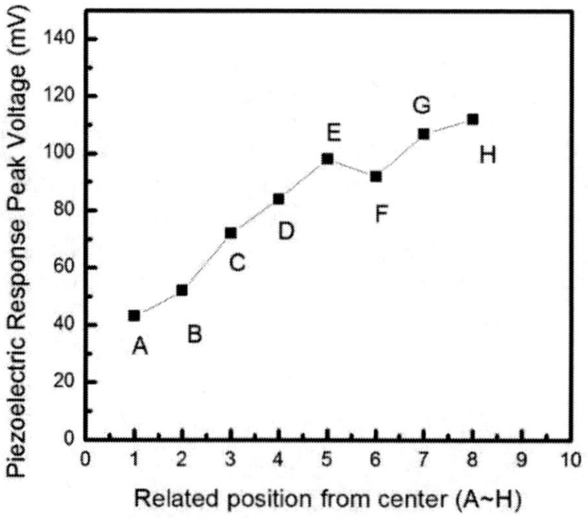

Figure 7. Peak voltage of piezoelectric response with related position

4 Conclusions

In this study, the ZnO thin films are deposited on SiNx/Si substrate. The crystalline characteristics of ZnO thin films are investigated through the XRD and SEM. The results show the ZnO thin film exhibits good crystalline pattern and surface morphology with controlled sputtering condition. The ZnO thin films sputtered using 2 inches target present various piezoelectric responses. With the exactly related position, a best piezoelectric response of ZnO thin film can be achieved.

Acknowledgment

The authors thank the Ministry of Science and Technology of the Republic of China, Taiwan, for financially supporting this research under contract numbers MOST 105-2632-E-366 -001.

References

1. R.C. Lin, Y.C. Chen, K.S. Kao, Appl. Phys. A Mater. Sci. Process. **89**,475 (2007)
2. M.-A. Dubois, P. Muralt, IEEE Trans. Ultrason., Ferroelect., Freq. Contr. **45,** 1169(1998)
3. T. Itoh, T. Suga, Appl. Phys. Lett. **64**, 37 (1994)
4. C. L. Wei, Y. C. Chen, K. S. Kao, K. T. Wu, D. L. Cheng, P. T. Hsieh, J. Micro/Nanolith. MEMS MOEMS, **9**, 031009(2010)
5. K. S. Kao, W. C. Shih, W. T. Ye, D. L. Cheng, Thin Solid Films, **605**, 77(2016)
6. J. Xia, S. Burns, M. Porter, T. Xue, G. Liu, R. Wyse, C. Thielen, IEEE Inter-national Frequency Symposium, 879 (1995)
7. W. Water, S.Y. Chu, Materials Letters, 55, 67 (2002)

The Impact of Iodoform on the Hydration, Bioactivity and Antimicrobial Properties of White Portland Cement

Nichola Jayne Coleman[1,a] and Qiu Li[2]

[1]*Faculty of Engineering and Science, University of Greenwich, Chatham Maritime, Kent, ME4 4TB, UK*
[2]*State Key Laboratory of Silicate Materials for Architectures, Wuhan University of Technology, Wuhan 430070, China*

Abstract. Iodoform (CHI_3) is a potential radiopacifying agent for use in Portland cement-based root-filling materials. During this study, the impact of 20 wt% iodoform on the hydration and setting of white Portland cement (WPC) was monitored by powder X-ray diffraction, ^{29}Si magic angle spinning nuclear magnetic resonance spectroscopy and Vicat apparatus. The presence of 20 wt% iodoform reduced the initial and final setting times of WPC from 150 to 121 min and 200 to 165 min, respectively. Iodoform had little impact on the products and extent of hydration after 7 days of curing; although, it did cause a reduction in the mean silicate chain length of the C-S-H gel (from 4.11 to 3.47 units). Both iodoform-blended and unblended cement pastes exhibited similar in vitro bioactivity, with the formation of crystalline hydroxyapatite on their surfaces within 1 day of exposure to simulated body fluid. An inhibition zone assay confirmed that WPC possesses intrinsic antimicrobial activity against *S. aureus*, *P. aeruginosa* and *E. coli*, which is significantly enhanced in the presence of iodoform. This study indicates that iodoform may be a suitable radiopacifying agent for Portland cement-based dental restoratives; although, further work is required to determine its long-term stability within the cement matrix.

1 Introduction

For the past two decades, Mineral trioxide aggregate (MTA), a commercial formulation comprising 80 wt% Portland cement and 20 wt% bismuth oxide (to confer radiopacity), has been used as a root filling material in dentistry [1, 2]. MTA is supplied as 1 g of powder which is manually mixed with 0.35 g of sterile water.

The major components of Portland cement are the impure phases; 'alite' (tricalcium silicate, Ca_3SiO_5), 'belite' (dicalcium silicate, Ca_2SiO_4), 'aluminate' (tricalcium aluminate, $Ca_3Al_2O_6$) and 'ferrite' (tetracalcium aluminoferrite, Ca_2 (Al/Fe) O5). Up to 5% of gypsum ($CaSO_4.2H_2O$) is also ground into the anhydrous cement during production to regulate the setting of the aluminate phase. 'Tooth-colored' MTA is formulated from white Portland cement (WPC) which contains low quantities of iron and manganese that impart the grey colour to ordinary Portland cement (OPC) from which 'Gray' MTA is produced [2].

The aluminate and ferrite phases react with water and gypsum to give rise to needle-like crystals of ettringite ($6CaO.Al_2O_3.3SO_3.32H_2O$) and its Fe-substituted analogue. Alite and belite react with water

[a] Corresponding author: n.coleman@gre.ac.uk

to form a poorly crystalline calcium silicate hydrate gel phase (C-S-H) of approximate formula $Ca_3Si_2O_7.3H_2O$ and portlandite (Ca (OH) $_2$). During hydration, the unpolymerised (Q^0) silicate tetrahedra of alite and belite become hydrated (Q^0 (H)) and then condense together to form dimers (Q^1). Ongoing condensation reactions produce short silicate chains comprising mid-chain (Q^2) species linked to two other silicate tetrahedra, and mid-chain Q^2(1Al) units linked to one silicate and one aluminate tetrahedron.

Iodoform (CHI_3) is among a range of candidate radiopacifying agents for use in Portland cement-based dental restoratives with potentially superior physicochemical, biological and antimicrobial properties to those of bismuth oxide [3-5]. Results of animal and human studies indicate that iodoform is sufficiently biocompatible and radiopaque at 20% addition to Portland cement [4, 5].

The objectives of this study were to investigate the impact of 20 wt% of iodoform on the hydration chemistry, in vitro bioactivity and antimicrobial properties of white Portland cement. Iodoform-blended (WPC-I) and unblended (WPC) cement paste samples were hydrated for 7 days prior to analysis by powder X-ray diffraction analysis (XRD), ^{29}Si magic angle spinning nuclear magnetic resonance spectroscopy (MAS NMR) and Fourier transform infrared spectroscopy (FTIR). The impact of iodoform on the initial and final setting times of the cement pastes was determined by penetration using a Vicat apparatus. The bioactivity of the cement samples was evaluated in vitro by monitoring the formation of hydroxyapatite on their surfaces, by FTIR, after immersion in simulated body fluid. And an inhibition zone assay was used to determine the antimicrobial activity of the cements against *Staphylococcus aureus*, *Pseudomonas aeruginosa* and *Escherichia coli*.

2 Materials and methods

2.1 Materials and sample preparation

The white Portland cement used in this study was supplied by Lafarge and is commercially available as 'Snowcrete' [6]. Iodoform, ex. Sigma Aldrich, UK, was used as received. Cement paste specimens (WPC) were prepared in duplicate by manually mixing the cement with distilled water at a water:cement ratio of 0.35 by mass. The samples were packed into polypropylene tubes, hermetically sealed and cured at 37 °C (i.e. body temperature). Samples blended with iodoform (WPC-I) were prepared similarly with partial replacement of the WPC by 20 wt% iodoform (at the same water:cement ratio).

2.2 Initial and final setting times

Initial and final setting times of the cement pastes were determined in triplicate in accordance with ASTM C191-08 using a manual Vicat apparatus [6]. Setting data were collected in triplicate and were subjected to a two-tailed *t*-test at $P = 0.05$.

2.3 Characterisation

For the purposes of characterisation, the specimens were cured for 7 hours before hydration was quenched by solvent exchange with propan-2-ol. Powder XRD was performed using a Philips D8 diffractometer with Cu Kα = 1.5406 Å, a step-size of 0.019° from 5° to 50° and a measuring time of 141.8 s per step.

MAS NMR spectra were collected on a JEOL JNM-ECX 300 MHz spectrometer. Single pulse ^{29}Si MAS NMR spectra were obtained with a pulse delay of 5 s, an acquisition time of 0.02048 s and 35000 scans. The ^{29}Si chemical shifts were referenced to tetramethylsilane (TMS) and the FIDs were processed by Delta software (provided by JEOL) to obtain spectra which were then analysed using Igor Pro software. The method used to deconvolute and analyse the ^{29}Si MAS NMR spectra is detailed elsewhere [6].

2.4 In vitro bioactivity analysis

For the analysis of bioactivity of WPC and WPC-I, cement pastes were cured for 7 days and immediately exposed to simulated body fluid (SBF) according to the method described by Kokubo and Takadama [7]. In each case, 0.15 g of cement sample were contacted with 150 cm^3 of SBF in hermetically sealed polypropylene containers at 37 °C for 3, 6, 24, 48, 72 and 168 hours. Each analysis was carried out in triplicate. The recovered cement samples were rinsed with deionised water, dried in air at 37 °C for 24 h and analysed by FTIR using a Perkin Elmer Paragon spectrometer.

2.5 Antimicrobial zone of inhibition assay

The antimicrobial properties of WPC and WPC-I were assessed using the inhibition zone method against *Staphylococcus aureus* NCIMB 9518, *Pseudomonas aeruginosa* NCIMB 8628 and *Escherichia coli* NCIMB 9132. Overnight cultures of each bacterium were spread on nutrient agar plates. Individual discs (1 cm diameter x 1 mm thickness) of cement were placed in the centre of each spread plate. Each assay was conducted in quadruplicate. The plates were examined for clear zones after incubation at 37 °C for 24 hours. The final population densities of the plates spread with *S. aureus*, *P. aeruginosa* and *E. coli* were approximately 2.4 x 10^8, 5.3 x 10^8, and 3.7 x 10^7 colony forming units per plate. The data were subjected to a one-tailed *t*-test at (n - 2) degrees of freedom and P = 0.05.

3 Results and discussion

3.1 Initial and final setting times

The initial and final setting times for samples WPC and WPC-I were obtained using a manual Vicat apparatus in accordance with ASTM C191-08, and are listed in Table 1. The presence of 20 wt% iodoform was found to reduce both the initial and final setting times of WPC by 29 min and 35 min, respectively.

Table 1. Initial and final setting times for WPC and WPC-I.

Cement	WPC	WPC-I
Initial set (min)	150 ± 15	121 ± 3
Final set (min)	200 ± 17	165 ± 5

3.2 Characterisation

Powder XRD data for WPC and WPC-I are presented in Fig. 1. The formation of ettringite and portlandite is noted in the diffraction patterns of both cement pastes, along with the residual unreacted parent cement phases, alite and belite [8]. The reflections of iodoform are seen to persist in the diffraction pattern of the bended paste up to 7 days, and there is no evidence for the formation of any other anomalous phases arising from the presence of iodoform within this cement system.

The subtracted, deconvoluted and fitted ^{29}Si MAS NMR spectra, together with the residue of the subtracted and fitted spectra, for the WPC and WPC-I pastes are shown in Fig. 2. These spectra comprise a sharp peak at -72 ppm which is assigned to the residual unreacted Q^0 species in belite, and hydrated Q^0(H), Q^1, Q^2(1Al) and Q^2 species of C-S-H gel in relative abundancies that are typical of white Portland cement that has hydrated for 7 days under similar conditions [6, 9]. The position and number of signals in the spectra of WPC are the same as those of WPC-I, and the extents of hydration of both cements (~60%) do not differ significantly (Fig. 2). This suggests that iodoform does not directly participate in the hydration reactions of white Portland cement; however, the decrease in

DOI: 10.1051/matecconf/201710904002

mean silicate chain length indicates that it may exert an influence on the assembly of the C-S-H gel phase.

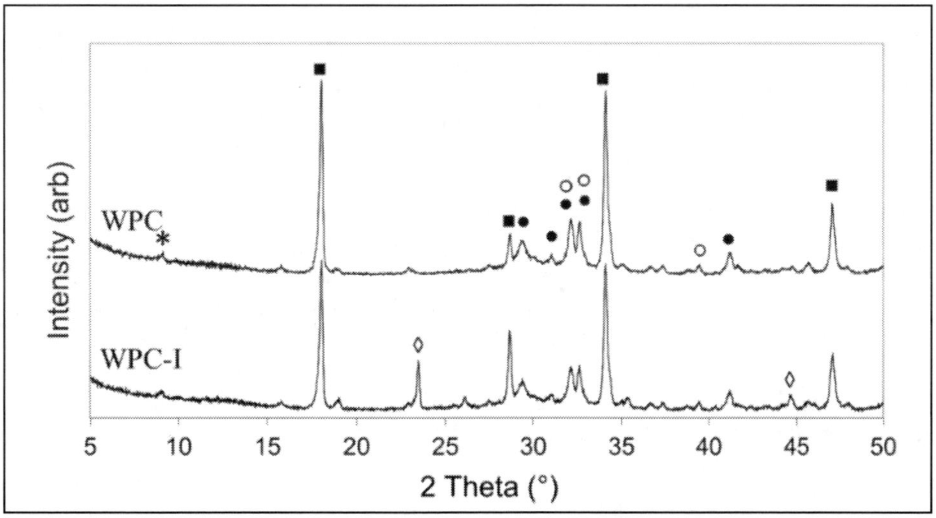

Figure 1. XRD data for WPC and WPC-I. (Key: ● alite; ○ belite; * ettringite; ■ portlandite; ◊ iodoform).

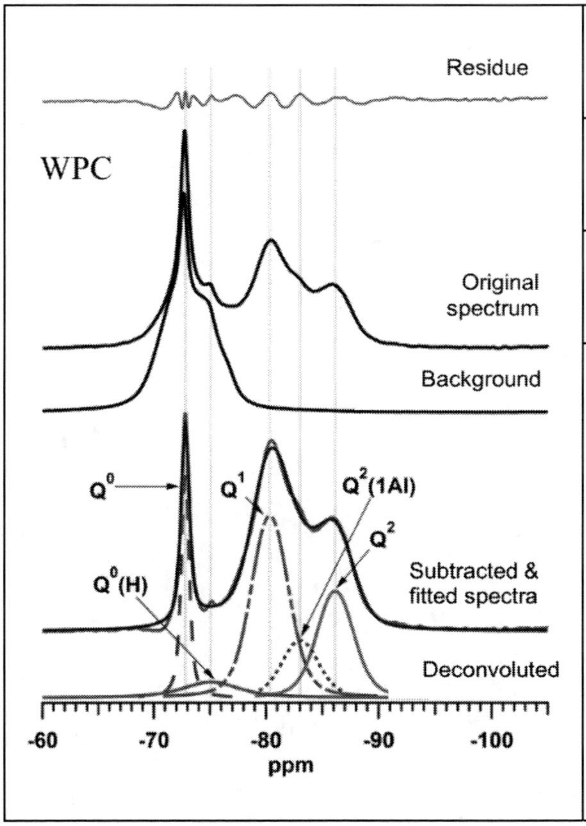

Sample	WPC	WPC-I
Extent of hydration	60%	61%
Mean silicate chain length	4.11	3.47

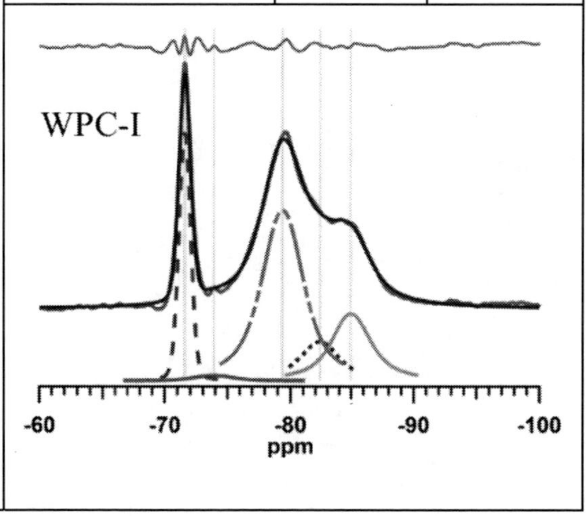

Figure 2. ^{29}Si MAS NMR data for WPC and WPC-I.

3.3 In vitro bioactivity analysis

The in vitro formation of a layer of bone-like substituted hydroxyapatite (HA), $Ca_{10}(PO_4)_6(OH)_2$, on the surface of a material placed in SBF solution provides an indication of its bioactivity (i.e. the ability of the material to bond with living bone tissue and to stimulate its regeneration) [7]. The FTIR spectra of WPC and WPC-I, as functions of residence time in SBF, are presented in Fig. 3. Si–O stretching modes of the C-S-H gel give rise to the broad band at ~970 cm^{-1} with contributions at 870 and 1120 cm^{-1} from carbonate and sulphate groups, respectively [8]. The unresolved doublet at 1420 and 1490 cm^{-1} also arises from carbonate species, and the signals at 1660 and 3470 cm^{-1} are attributed to O–H vibrations of water and hydroxyl groups. Stretching modes of O–H in portlandite are assigned to the sharp signal at 3670 cm^{-1}. Weak sharp bands arising from iodoform are noted to contribute to the spectrum of WPC-I at 2975, 1263 and 868 cm^{-1}.

The broad signal at 600 cm^{-1} in the FTIR spectra of the cements exposed to SBF for 1 hour, is indicative of the rapid deposition of a layer of amorphous hydroxyapatite on their surfaces (Fig. 3) [8]. The increase in relative intensity and resolution of this band into a doublet signify the ongoing development and crystallisation of the hydroxyapatite layer as a function of residence time in SBF. The concurrent disappearance of the sharp O–H signal at 3670 cm^{-1} in the FTIR spectra of both cements after 1 hour in SBF, is attributed to the dissolution of portlandite. These results confirm that the in vitro bioactivities of WPC and WPC-I are similar and that the incorporation of iodoform into the cement mixture does not adversely affect this clinically important property of the material.

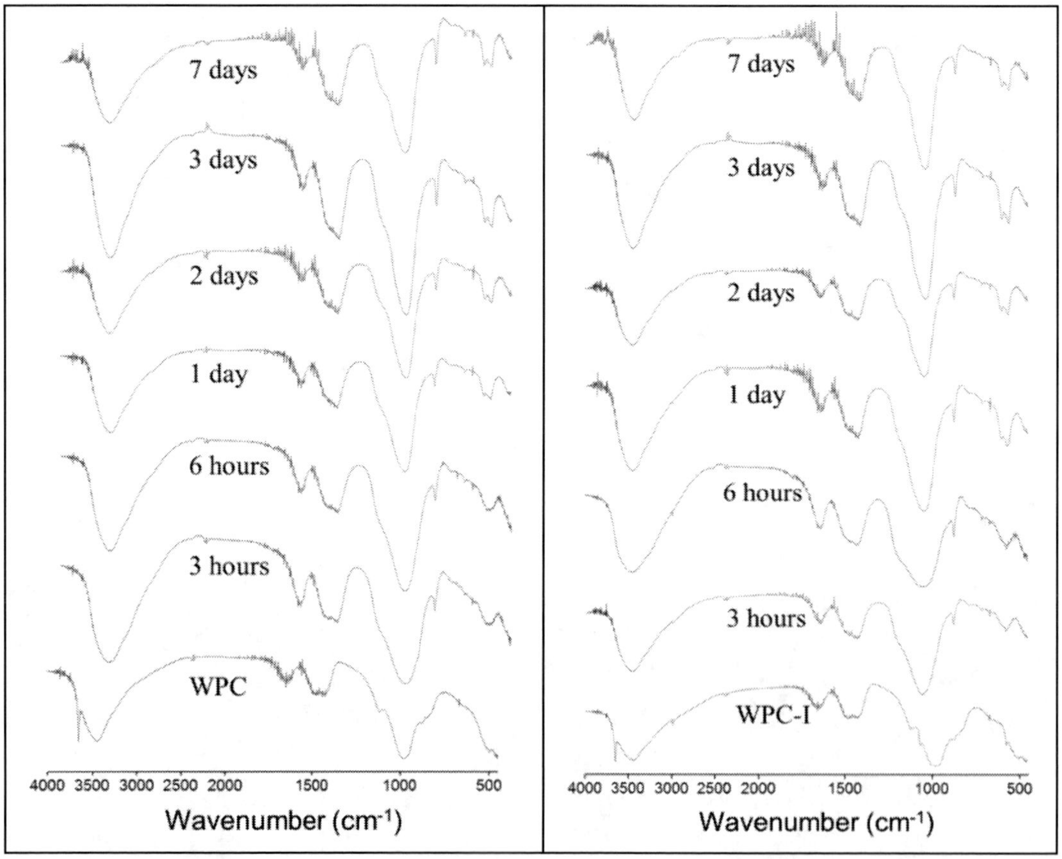

Figure 3. FTIR spectra of WPC and WPC-I as functions of residence time in SBF.

3.4 Antimicrobial zone of inhibition assay

The results of the antibacterial inhibition zone assays using *S. aureus*, *P. aeruginosa* and *E. coli* are listed in Table 2. Clear zones around each of the WPC discs were noted in all cases indicating that this material is intrinsically antimicrobial against all bacterial strains. There were found to be significant increases in the size of the inhibition zones around the WCP-I discs for all bacteria (P = 0.05), indicating that the incorporation of iodoform in the cement paste enhances its antimicrobial properties with respect to both Gram-positive (*S. aureus*) and Gram-negative (*P. aeruginosa*, *E. coli*) bacteria.

Table 2. Inhibition zone data for WPC and WPC-I.

Bacterium	Diameter of zone of inhibition (cm)	
	WPC	WPC-I
S. aureus	16.8 ± 1.5	18.0 ± 0.1
P. aeruginosa	14.5± 0.9	16.8 ± 1.5
E. coli	18.8 ± 1.0	20 ± 0.1

3.5 The role of iodoform as a radiopacifier in Portland cements

Iodoform is among the candidate radiopacifying agents that are currently considered for incorporation in Portland cements for dental applications [3-5]. The present study has demonstrated that 20 wt% replacement of iodoform for white Portland cement reduces both initial and final setting times. The ability of a radiopacifying agent to reduce setting time is a distinct benefit, as a common complaint regarding MTA is that its setting time is inconveniently long [10]. Bismuth oxide, the proprietary radiopacifier in MTA, is known to prolong both the initial and final setting times [11], so in this respect, iodoform would be an advantageous alternative.

This study has also confirmed that iodoform does not directly participate in or divert the normal hydration reaction of the cement; however, the mean silicate chain length of the C-S-H gel was reduced in the presence of iodoform, indicating that it may play a structure-directing role in the cement system. Further work to explore this phenomenon, and also to investigate the impact of iodoform on the evolution of the calcium aluminate hydrate phases is now warranted. Since the long-term chemical stability of iodoform in the highly alkaline environment of the cement matrix has not yet been established, further investigations of this nature are also recommended.

Additionally, the incorporation of iodoform into the cement system was not found to have an adverse impact on bioactivity in vitro. A layer of bone-like HA precipitates on the surface of an implanted bioactive material in vivo (from the components of human plasma) which provides a focus for bone-forming cells to adhere and proliferate. This is an acknowledged advantage of Portland-cement based biomaterials that exhibit the ability to stimulate the formation of new cementum, pulp and bone [3]. Animal and human studies have indicated that 20:80 mixtures of iodoform and Portland cement exhibit similar biocompatibility to that of commercial MTA [4, 5]. Hence, from a biological perspective, iodoform has the potential to enhance the antimicrobial properties of the cement without compromising its bioactivity and biocompatibility.

4 Conclusions

This study considers the impact of 20 wt% iodoform, a radiopacifying agent, on the hydration and setting of white Portland cement (WPC) by powder X-ray diffraction analysis, ^{29}Si magic angle spinning nuclear magnetic resonance spectroscopy, Fourier transform infrared spectroscopy and penetration by Vicat apparatus. The presence of iodoform was found to reduce the initial and final setting times of WPC by 29 min and 35 min, respectively. Iodoform had no significant impact on the products and extent of hydration after 7 days; although, it did cause a reduction in the mean silicate chain length of the C-S-H gel (from 4.11 to 3.47 units). Both iodoform-blended and unblended cement pastes exhibited similar in vitro bioactivity, with the formation of crystalline hydroxyapatite on their surfaces within 1 day of exposure to simulated body fluid. An inhibition zone assay confirmed that WPC possesses intrinsic antimicrobial activity against *S. aureus*, *P. aeruginosa* and *E. coli*, which is significantly enhanced in the presence of iodoform. This study confirms that iodoform may be a suitable radiopacifying agent for Portland cement-based dental restoratives; although, further work is required to determine its long-term stability within the cement matrix.

References

1. N.J. Coleman, J.W. Nicholson, K. Awosanya, Cement Concr. Res. **37**, 1518 (2004).
2. Q. Li, N.J. Coleman, Dent. Mater. J. **34**, 458 (2015).
3. E.A. Bortoluzzi, J.M. Guerreiro-Tanomaru, M. Tanomaru-Filho, M.A.H. Duarte, Oral Surg. Oral Med. Oral Pathol. Oral Radiol. Endod. **108**, 628 (2009).
4. N. Marques, N. Lourenço Neto, A.P. Fernandes, C. Rodini, M. Hungaro Duarte, D. Rios, M.A. Machado, T. Oliviera, J. Microscopy **260**, 281 (2015).
5. N. Lourenço Neto, N.C.T. Marques, A.P. Fernandes, C.O. Rodini, M.A.H. Duarte, M.C. Lima, M.A.A.M. Machado, R.C.C. Abdo, T.M. Oliviera, J. Oral. Sci. **56**, 29 (2014).
6. N.J. Coleman, Q. Li, Mater. Sci. Eng. C, **33**, 427, (2013).
7. T. Kokubo, H. Takadama, Biomaterials **27**, 2907 (2006) .
8. N.J. Coleman, J.W. Nicholson, K. Awosanya, Cement Concr. Res. **37**, 1518 (2004) .
9. Q. Li, A.D. Deacon, N.J. Coleman, Dent. Mater. J. **32**, 808 (2013)
10. B. Czarnecka, N.J. Coleman, H. Shaw, J.W. Nicholson, Dent. Med. Probl. **45**, 5 (2008).
11. K.S. Coomaraswamy, P.J. Lumley, M.P. Hofmann, J. Endod. **33**, 295 (2007) .

MATEC Web of Conferences **109**, 04003 (2017) DOI: 10.1051/matecconf/201710904003
ICMSNT 2017

Effect of Acclimatization Time to Microbial Cell Growth and Biosynthesis of Mesophilic Gammaproteobacterium, in Orbital Shake Flasks

Abd. Aziz Mohd Azoddein[1], Mani Malam Ahmad[1,a], Rosli Mohd Yunus[1] and Nik Meriam Nik Sulaiman[2]

[1]*Faculty of Chemical and Natural Resources Engineering, University Malaysia Pahang, Lebuhraya Tun Razak, 26300 Gambang, Kuantan, Pahang, Malaysia*
[2]*Department of Chemical Engineering, Faculty of Engineering, University Malaya, Kuala Lumpur, Malaysia*

Abstract. Growth pattern of *Pseudomonas putida* (ATCC 49128), was found to predominantly rely on the age of the inoculums, prior to its contact with physical and chemical agents and nutrient availability. Under suitable inoculums, bacteria tend to grow faster in a batch type of growth pattern which is usually sustained until after nutrient depletion. In this research, the bacterial growth pattern was studied in an incubator shake flask using 8 g nutrient media and physical operational parameters temperature of 37 °C and agitation of 180 rpm over a period of 24, 48 and 72 hours. Prior to this, *P. putida* was added into 20.0 ml nutrient broth and incubated in an incubator for 24 hours at 37 °C, before adding it to 180 ml nutrient broth 30% (v/v^{1-}). Growth, via acclimatization was initially observed after 1hr of inoculation with an overwhelming exponential growth of 2.69-2.57 within first 24 hr, exceeding the 48 and 72 hrs ranges. This additionally relates to particular cell biomass growth rate (μ) of 0.58 hr^{1-}, 3.87 number of generation (n), generation time (*g*) 1.09 and growth rate constant (*k*) of 0.01 hr^{1-}, achievable within 24 hrs. It was therefore concluded that the sensitivity of this strain to time is significant, as optimal growth was achieved within 24 hrs of acclimatization, thereby showing a drastic reduction in the time of growth.

1 Introduction

Metabolism refers to the concept of biochemical activities that enable the organisms grow and reproduce effectively in an appropriate substrate. This includes its utilization for cell repair and biosynthesis as well as well other necessary physiological transformations associated with this [1]. These biochemical activities involve the brake down of substrate through dissimilation to release energy, as well as assimilation of organic and inorganic substrates for growth and biosynthesis of bacterial cells. Growth in microbes is the most essential response to their physiochemical environment [2].

Upon inoculation into a suitable media, this gram negative, aerobic and gammaproteo bacterium selectively takes up nutrients from the medium for biosynthesis [3]. The initial lag growth phase is a period of cell change in accordance with the new environment. The increase in cell is very minimal and is dependent upon cells' former history (age, pre-exposure to damaging and toxic substances and

[a] Corresponding author: mmahmadu@gmail.com

the medium). In commercial production of enzyme, the initial growth stage used to be a brief as possible, depending on the type of inoculum used [4].This is then followed by exponential or log, within which the isolate have modified to the new medium. Thereafter these cells now divide aggressively which usually occurs at 15-20 minutes rate, with a balanced cell biomass production phase [5]. This is the peak nutrient utilization stage and growth rate is directly proportional to nutrient concentration in the medium During the last stage refer to as stationary or death phase, growth rate is equal to the death phase, thus termed as zero growth rates and is characterized by nutrient exhaustion and metabolites accumulation [6]. Despite this zero growth, yet the cells retained active metabolic activities. Ideally, growth in bacterial cells follows a geometric progression pattern (i.e. 2^0! 2^1! 2^2! 2^n) which is depicted using Monod growth model [7].

$$X_t = X_{0}e^{ut} \tag{1}$$

where μ is microbial biomass increase or specific growth rate (h^{r-1}), X_t, cell growth over a period of time t, and X_0 initial cells number. Taking log of the two sides, the above model equation is given as

$$\log X_t = \log X_0 + \mu t \tag{2}$$

Specific growth rate, μ which is addition of cell biomass per unit time, is given;

$$\mu = \frac{\ln X_t - \ln X_0}{t} \tag{3}$$

Number of generation, n

$$n = 3.3 \ln \frac{X_t}{X_0} = 3.3 \ln \frac{OD}{OD_0} \tag{4}$$

Generation time, g time required for cells to doubled

$$g = \frac{t}{n} \tag{5}$$

where t is the period of active cell growth given in days, hours or minutes, *n* number of generation The number of generation at an active cell growth phase is termed as growth rate constant *k*.

$$k = \ln \frac{2}{g}$$

This piece of work is envisage at evaluating the growth pattern of *P. putida* over a specified period upon subjecting it to single nutrient medium and physical operational parameters of an orbital shake flasks. Acclimatization time is an important aspect to microbiologist, especially as it applies to industrial and biotechnological application in fermentation, bioremediation, enzyme production among others. Therefore, the finding will be of significant value when utilizing this isolates in any experimental and real application.

2 Methodologies

2.1 Strain and Cultivation

The *P. putida* (ATCC 49128) with its growth media was obtained from a licensed supplier in USA through a local agent Merck (Malaysia) Sdn. Bhd. Culture media was formulated by required amount of nutrient in a liter of water and agitated vigorously with magnetic stirrer on a hot plate until it dissolved. The solution was then autoclave at 121°C for 15 minutes; and to cool to a temperature of less than 40°C inside a safety cabinet before transferring to individual 20 ml sampling bottles. Seeding was done by suspending 1-3 loops of the inoculum for effective viability, into the previously prepared media in 20 ml bottles 10% (wv^{1-}). The culture was then incubated at 37°C for one day inside a microbiological incubator. Thereafter, the isolate was emptied into a 150 ml nutrient broth 30% (vv^{1-}) [8], and placed into the orbital shaker. These processes were done aseptically inside a biosafety cabinet, as well in according to standard. It was indicated that multiple loops seeding ensure proper bacterial growth [3]. The orbital shaker was operated at 37°C and 180 rpm. The experiment was conducted in three different acclimatization times of 24, 48 and 72 hours, respectively.

2.2 Analysis of sample for different *P. putida* growth parameters

For growth determination, 2.5 ml Aliquots samples were withdrawn periodically using 5 ml syringe aseptically. Growth was measured using U-VIS Spectrophotometer (Hatachi, U-1800), at $OD_{600\ nm}$ wavelength. This mechanism was based on the principle of light diffraction, reflection and absorption by suspended cells. This technique provides fast, easy, inexpensive and reliable for estimating cell density. The magnitude of transmitted light is proportional to cell density or biomass regulated by media absorption and scattering. While for cell dry weight measurement, 1 ml aliquot sample was collected in a centrifuge vails and spin at 12,000 rpm for 15 min. The floating liquid was disposed and the sediment washed in a 5% saline solution and centrifuged again, then oven dried at 100°C- 110°C for 1hr. The mass was measured and dried repeatedly until a stable weight was obtained [9].

3 Results and Discussion

3.1 Effect of Acclimatization time of *P. putida* Growth

Result from Figure 1, Table 1, shows that the cultured *P. putida* has been acclimatized within 24 hours and adapted well to its new environment of shake flask. Meanwhile, it took this strain just an hour to adapt and start growing (lag phase) after inoculation. The cell optical density (OD), number of generation, *n* and growth rate constant, *k* decreased to zero in the lag phase. In this condition, *P. putida* is considered to be in non-productive period and more time needed to adapt to the new environments. During this period, molecular components were usually re-constituted largely due to slight increase in cell concentration. Thereafter, growth rate increases in the subsequent 2 to 5 hours and as a result the cell density also increased. Consequently, the OD and the exponential cell growth were at their maximum values of 2.57 and 2.70 within 24 hours acclimatization time, respectively. This period correspond to the rapid production of new cell measured as biomass. However, growth was maintained and observed to decelerate gradually thereafter. This corresponds to a slight upward shift of the medium pH (Tab. 1), thus drastic increase in metabolic rates of the isolate. Increase in cell biomass (μ) of 0.58 hr^{-1} in 24, 0.56 hr^{-1} in 48 and 0.55 hr^{-1} in 72 hours acclimatization time, respectively. Specific growth rate, (μ) or cell biomass increase is used to indicate the average growth proportion of the entire cells in a culture medium [10].

Based on the analysis, the maximum number of generation, n were 3.87 for 24 hours acclimatization, 3.57 for 48 hours acclimatization and 3.61 for 72 hours acclimatization. While the calculated maximum generation time, g were 14.00 hours for 24 hours, 2.00 hours for both 48 hours and 72 hours acclimatization. Also, the growth rate constant, k estimated were 0.01 hr-1, 60 hr-1 and 0.66 hr-1 for 24, 48, and 72 hours, respectively. Based on aforementioned growth parameters recorded, it was quite glaring that, this isolate require a minimum of 24 hours to acclimatize and start to grow effectively in its new environment.

Table 1. Summary of parameters related to *P. putida* growth for 24 h, 48 and 72 hours acclimatization time

Growth Parameter Time (hr)	Min			Max		
	24	48	72	24	48	72
Proportionate cell growth (μ)	0.58	0.56	0.55	-	-	-
OD (600 nm)	0.17	0.17	0.17	2.57	2.09	2.14
Exponential cell growth (ln OD/OD_o)	0.05	1.79	2	2.7	2.49	2.52
Number of Generation *n*	0.07	2.57	2.86	3.87	3.57	3.61
Generation Time *g* (hr)	1.09	1.17	1.05	1.76	1.68	1.66
Growth rate Constant, *k*	0.05	0.42	0.42	0.63	0.6	0.11
pH	6.24	6.59	6.6	7.46	7.42	7.47

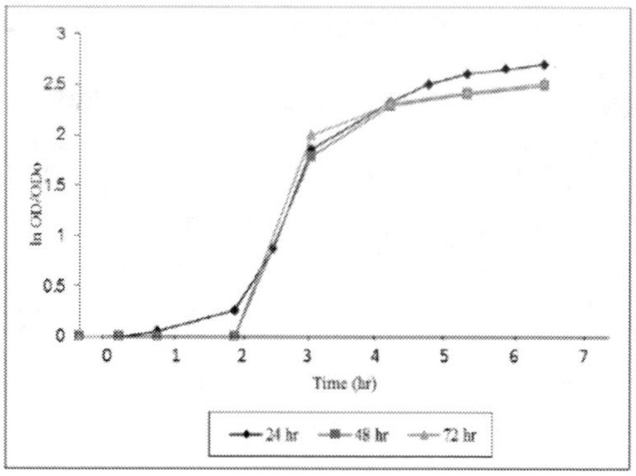

Figure 1. Exponential cell growth of *P. putida* within first 6 hours of inoculation

3.2 Determination of *P. putida* Growth Parameters

The results showed a drastic increased in *P. putida* growth within the first six hours of inoculation with corresponding OD range of 0.21 to 1.98. This can be seen as per average number of generation 3.19 and the maximum growth rate constant value of 0.37 hr^{1-}. It was reported that based on the growth parameters, different microorganisms used to grow and acclimatized under the distinct microbial media [6]. This is significant for optimizing nutrient composition as well as determining its effects on the bacteria culture (Table 2). Based on the findings from this study, it can be seen that *P. putida* growth pattern could fully be understood from experimentation in an enclosed system of orbital shake flask, which is considered a batch culture.

Table 2: Different growth parameters related to *P. putida* growth over 30 hours period

Time (hr)	OD	Exponential cell growth (ln OD/OD_O)	Number of generation (*n*)	Generation time (*g*)	Growth Rate Constant (*k*)
2.00	0.21	5.36	0.00	-	-
4.00	0.68	6.53	1.67	2.40	0.29
6.00	1.98	7.59	3.19	1.88	0.37
7.00	2.10	7.65	3.28	2.14	0.32
8.00	2.25	772	3.38	2.37	0.29
10.00	2.36	7.77	3.44	2.90	0.24
11.00	2.39	7.78	3.46	3.18	0.22
26.00	2.59	7.86	3.58	7.83	0.09
28.00	2.59	7.86	3.58	7.83	0.09
30.00	2.57	7.85	3.57	8.41	0.08

During the first 6 hours of the 30 hour experiment the isolate attained a maximum growth of 7.86. After 6 hours, it was noted that the pattern of cell concentration is consistent and decreased slightly with time and an equilibrium growth rate was recorded. The proportional cell biomass increase (μ) was found to be 0.96 hr^{-1}. It was reported that growth is limited mainly by factors such as temperature, substrates concentration and composition as well as genomic of the organism itself [5]. Number of generation, *n* for *P. putida* was 1.67 to 3.58, while the generation time, *g,* which is the time interval between two cell divisions, is from 1.88 to 8.41 hours. The growth rate constant, *k* is in the range of 0.08 to 0.37 hr^{-1}, which was a measure of how fast the cells are dividing in a culture for the first 2 to 6 hours from the 30 hours experiments. The result obtained agreed with the previous findings where *P. putida* was found to overwhelmingly grow in a recalcitrant medium containing toxic organic wastes [11].

3.3 Determination of Biomass Synthesis Using Cell Dried Weight

Figure 2, presented the finding for *P. putida* (ATCC 49128) cell dried weight (gl^{1-}). To better understand the nature of this growth pattern, a calibration curve of OD was plotted against cell dry-weight, although the curve tend to be non-linear as OD approaches 0.30 [6]. The result shows that cell dried weight increase was directly proportional to optical density. Increase in OD is based on the principle of rate of light absorbance passing through a medium containing cell biomass. Consequently as the cell growth increase due to biosynthesis of cell biomass the less the amount of light passing through and the higher the rate of light absorbed on to the cell surface.

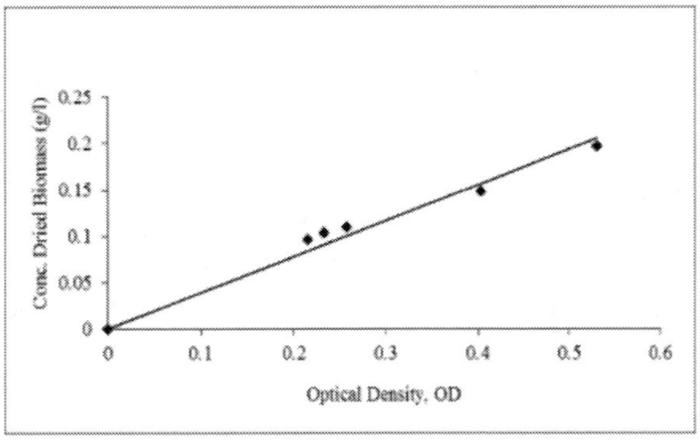

Figure 2. *Pseudomonas putida* cell dry weight curve

4 Conclusions

In conclusion, it was obvious that P. putida (ATCC 49128) bacterial isolate could be cultivated in a simple media composition under physical operational parameters of orbital shake flasks within a shortest period of less than 24 hours. Acclimatization indeed was found to be affected by many factors such as age of the culture and stress phenomenon or growth limiting factors. Growth was exponentially recorded within the first six hours of inoculation with lag phase lasting not more than an hour. Based on the condition applied in the medium, the P. putida growth usually kick-start only after a short period of time of inoculation into fresh medium. This period may be shorten or extended, but rely on the nutrient composition, genetic composition and age of the culture as well. However, generation time was found to be extended if the culture is taken from an old culture and transferred into fresh medium, even though the cells in the culture are viable. Results from this study indicated that P. putida, requires a minimum of 24 hr to adapt and grow optimally when inoculated into a fresh medium, though significant values were also observed for 48 and 72 hours, respectively. This further substantiated the significant role of acclimatization time on growth and biosynthesis of this isolate.

Acknowledgement

We wish to sincerely register our appreciation to University Malaysia Pahang, for their financial support through local research grant.

References

1. Zinkernagel, K.F.H.K.A.B.J.E. and R.M., 2005. *Medical Microbiology* 5th Edition, ed., Stutgart: Thieme. Available at: http://pgim.srilanka.healthrepository.org/handle/123456789/9873.
2. Volesky, B. (2007). Biosorption and me. *Water Research*, *41*(18), 4017–4029. http://doi.org/10.1016/j.watres.2007.05.062
3. Fonseca, P., Moreno, R., & Rojo, F. (2011). Growth of Pseudomonas putida at low temperature: Global transcriptomic and proteomic analyses. *Environmental Microbiology Reports*, *3*(3), 329–339. http://doi.org/10.1111/j.1758-2229.2010.00229.x
4. Rashedi, H., Izadi, A., Bidhendi, M. E., & Rashedi, H. (2015). Optimization of Operational Parameters in Rhamnolipid Production by Pseudomonas aeruginosa MM1011 in a Miniaturized Shaken Bioreactor, *2*(3), 271–278.
5. Shea, M., Litvin, S., & Chirnside, A. (2013). The Effect of Nitrogen , Sulfur , and Phosphorus Compounds on Bioremediation of Oil Spills by Pseudomonas fluorescens and Bacillus subtilis, 30–37
6. Shuler, M.L. and Kargi, F. 2002. Bioprocess Engineering. Sec. Ed. Prentice Hall PTR, Prentice-H all, Inc. Upper Saddle River, NJ
7. Bitton, G. (2001). *Wasterwater microbiology* (Vol. 40). http://doi.org/10.1002/1521-3773(20010316)40:6<9823::AID-ANIE9823>3.3.CO;2-C
8. Standbury, P.F., Whitaker, A. and Hall, S. J. (1984). P. of F. T. O. B. H. (1984). *Principle of Fermentation. Standbury, P.F., Whitaker, A. and Hall, S. J. (1984). Principles of Fermentation Technology. Oxford: Butterworth Heinemann* (2nd ed., Vol. 53). OXFORD: BUTTERWORTH HEINEMANN. http://doi.org/10.1017/CBO9781107415324.004
9. Bratbak, G., & Dundas, I. (1984). Bacterial dry matter content and biomass estimates. *Applied and Environmental Microbiology*, *48*(4), 755–757.
10. Lee, K., Park, J. W., & Ahn, I. S. (2003). Effect of additional carbon source on naphthalene biodegradation by Pseudomonas putida G7. *Journal of Hazardous Materials*, *105*(1–3), 157–167. http://doi.org/10.1016/j.jhazmat.2003.08.005
11. Abuhamed, T., Bayraktar, E., Mehmetoğlu, T., & Mehmetoğlu, Ü.̈ (2004). Kinetics model for growth of Pseudomonas putida F1 during benzene, toluene and phenol biodegradation. *Process Biochemistry*, *39*(8), 983–988. http://doi.org/10.1016/S0032-9592(03)00210-3

Magneto Caloric Properties of Polycrystalline Gd_2O_2S for an Adiabatic Demagnetization Refrigerator

H.Fukuda[1,a], Y.Kobayashi[1], R.Arai[1], K.Baba[1], H.Nakagome[1] and T. Numazawa[2]

[1]*Department of Urban Environment System, Chiba University Chiba 263-8522, Japan*
[2]*Environment and Energy Materials Division, National Institute for Materials Science Tsukuba, Ibaraki 305-0003, Japan*

Abstract. Currently, many space missions that use cryogenic equipment are being planned. In particular, high resolution sensors, such as transition edge sensors, require very low operating temperatures, below 100 mK. Adiabatic demagnetization refrigerator (ADR) systems are a useful tool for producing ultra-low temperatures in space because these devices can operate independently of gravity. The magnetic material is one of the most important components with respect to effectiveness of the cooling power. Thus, we could increase the cooling power using a magnetic material that has a large entropy change over the operating temperature range. Polycrystalline Gd_2O_2S (GOS), which was developed by Numazawa et al, can be used as such as a magnetic regenerator material. Furthermore, GOS has a very large specific heat and a magnetic phase transition temperature of about 5.2 K. These features make GOS suitable for use in the high temperature stage of an ADR. In this study, we measured and evaluated the physical properties of GOS for applications to ADRs.

1 Introduction

Many scientific measurements have been performed in microgravity environments in recent years. In the study of astrophysical phenomena the polarimetry of cosmic microwave background (CMB) is a particularly active field of research.

Detectors used in polarimetry of CMB are transition edge sensor (TES) type microcalorimeters. A TES type microcalorimeter requires an ultra-low temperature operating environment of less than 100 mK [1]. Thus, it is necessary to develop refrigerators that operate below 100 mK and can be used in space. To produce temperatures below 100 mK in space, Shirron et al. devised the continuous adiabatic demagnetization refrigerator (CADR) [2] [3]. We have also studied a continuous adiabatic demagnetization refrigerator (ADR) system consisting of four stages [4].

The ADR uses a magnetocaloric effect [5] that is caused by an entropy change of a magnetic material when the external magnetic field is changed. The entropy change of the magnetic material is described by equation (1), derived from Maxwell relations. (M : magnetization, H : magnetic field)

$$dS = \left(\frac{\partial M}{\partial T}\right)_H dH \qquad (1)$$

From equation (1), the entropy change of a magnetic material can be determined by integrating the differential of the magnetization of the magnetic material in the external magnetic field. Thus, a large entropy change can be obtained if the differential of the magnetization of the magnetic material is

[a] Corresponding author: fukuda393@gmail.com

large. Thus, magnetic materials that have large entropy changes in desired temperature range could be used to increase the cooling power of ADRs.

Polycrystalline Gd_2O_2S (GOS), which was developed by Numazawa et al [6] [7], be used as such a magnetic regenerator material. Moreover, GOS is suitable for use in the high temperature stage of an ADR because its magnetic phase transition temperature is about 5.2 K. In this study, we measured the magnetization and specific heat of GOS and evaluated the physical properties of GOS for application to ADRs.

2 Polycrystalline Gd2O2S

GOS is an antiferromagnetic material that has its spins ordered in an antiparallel arrangement. Figure 1 illustrates an ordered arrangement of ferromagnetic and antiferromagnetic spins. A ferromagnetic material has a spontaneous magnetic moment that exists even if no magnetic field is applied. Antiferromagnetic materials do not have a spontaneous magnetic moment because their spins are ordered in an antiparallel arrangement. Moreover, antiferromagnetic materials generally do not feature strong magnetization when magnetized by external magnetic fields because their spins cancel each other. However, when the external magnetic field becomes strong the spins become aligned in the same direction as the external magnetic field. This phenomenon is called spin-flop and it produces a considerable magnetization change when all the spins become aligned in the same direction. If spin-flop is applied to an ADR, it is possible to induce large temperature changes because the large magnetization change leads to a large entropy change. GOS has been put to practical use as a magnetic regenerator material because it typically shows a λ phase transition and has a huge specific heat.

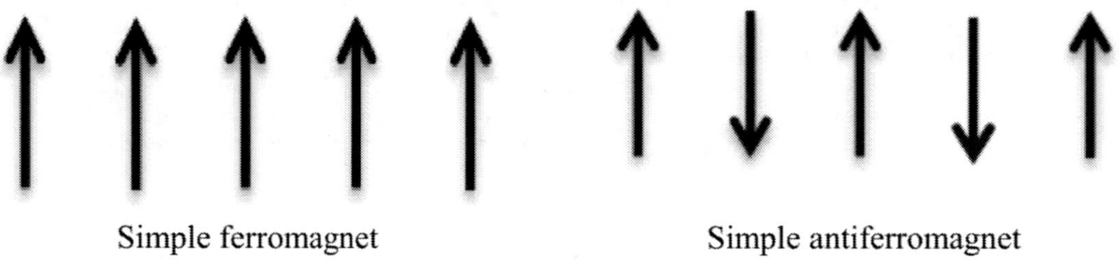

Figure 1. Ordered arrangement of spins

3 Experimental results

We measured the magnetization of GOS using a magnetic property measurement system (MPMS, Quantum Design) and the specific heat of GOS using a physical property measurement system (PPMS, Quantum Design). Figure 2 shows the sample of GOS used in measurement. The magnetization of GOS in an external magnetic field was measured at 4, 5, and 6 K when magnetized up to 5 T. The variation of GOS magnetization with temperature was measured at 0.1, 3, and 5 T for a temperature increase from 2.2 to 15 K. The change in specific heat with temperature was measured at 0, and 5 T for an increase from 2.2 to 15 K. The variation of GOS magnetization in an external magnetic field, the variation of GOS magnetization with temperature, and the specific heat with temperature are shown in Figures 3–5, respectively.

A spin–flop occurring in these systems should appear as a large magnetization change. However, Figure 3 shows that the magnetization of GOS increased linearly indicating that no spin-flop occurred. One reason for this is that GOS requires a magnetization greater than 5 T to reverse its spins.

Next, we focus on Figure 5, the specific heat with temperature. From this result, we determined the phase transition of GOS to be around 5.2 K. In addition, we calculated the entropy change of GOS using equation (2), as show in Figure 6, and interpolated the entropy using external interpolation.

DOI: 10.1051/matecconf/201710904004

$$S = \int_0^T \frac{C}{T} dT \tag{2}$$

Figure 6 shows that there was no effective entropy change below 4 K, which is the high temperature stage of an ADR. The entropy change of GOS was less than that of $GdLiF_4$, which is currently used in the high temperature stage of ADRs [8]. However, when we see figure 7 and 8, we could confirm that the entropy curve at 0 T was larger than that at 5 T, between 5 and 6 K. This is considered an inverse magnetocaloric effect [9] [10] [11]. The entropy of GOS and the entropy of GOS between 5 and 6 K are shown in Figure 7 and 8.

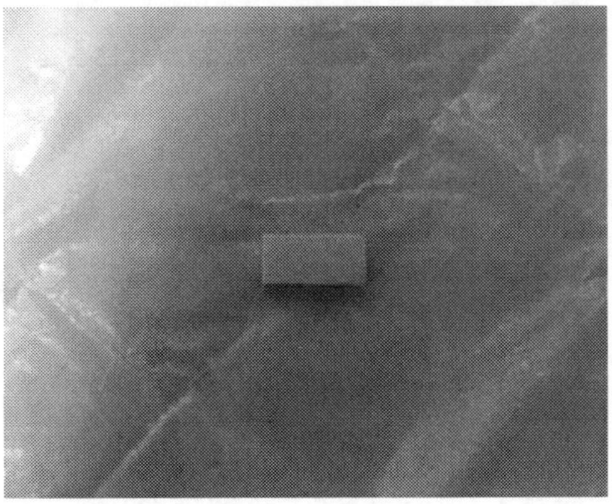

Figure 2. Sample of GOS used in measurement

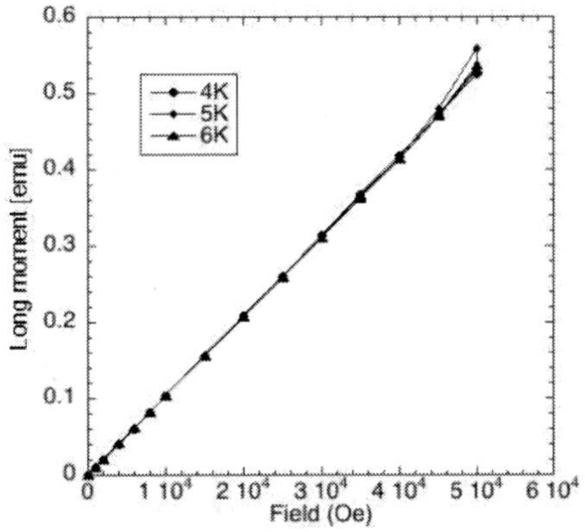

Figure 3. Variation of Gd2O2S magnetization with external magnetic field at 4, 5, and 6 K when magnetized to 5 T.

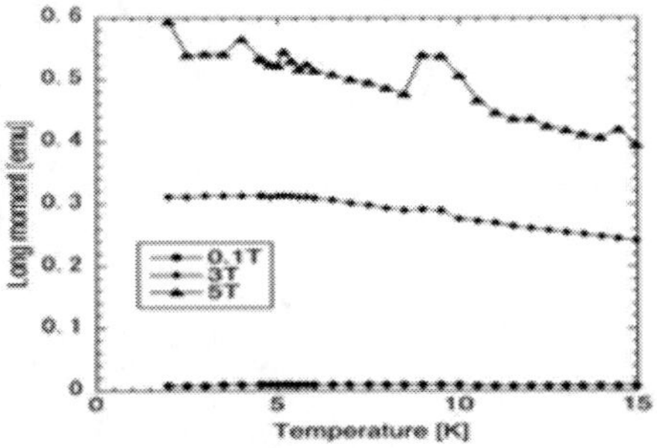

Figure 4. Variation of Gd2O2S magnetization with temperature at 0.1, 3, and 5 T as temperature was increased from 2 to 15 K.

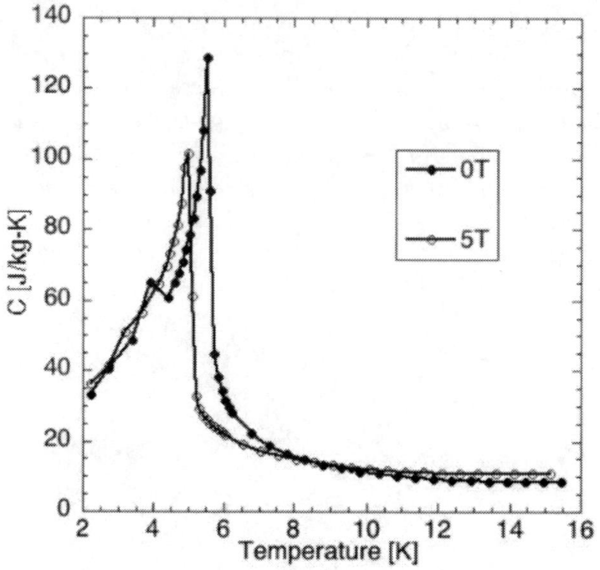

Figure 5. Specific heat with temperature at 0, and 5 T when temperature was increased from 2.2 to 15 K .

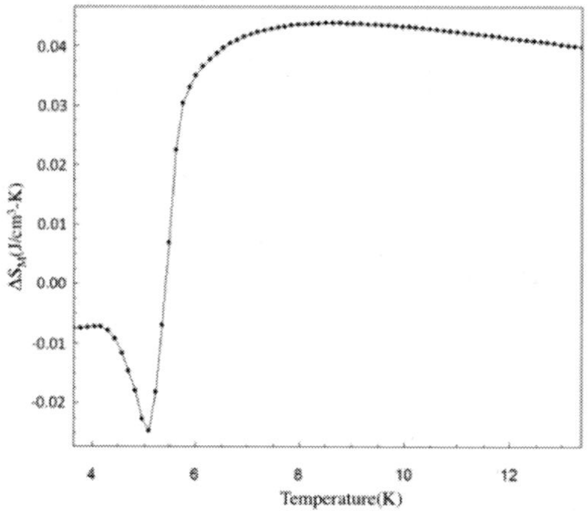

Figure 6. Entropy change of Gd_2O_2S as functions of magnetic field and temperature.

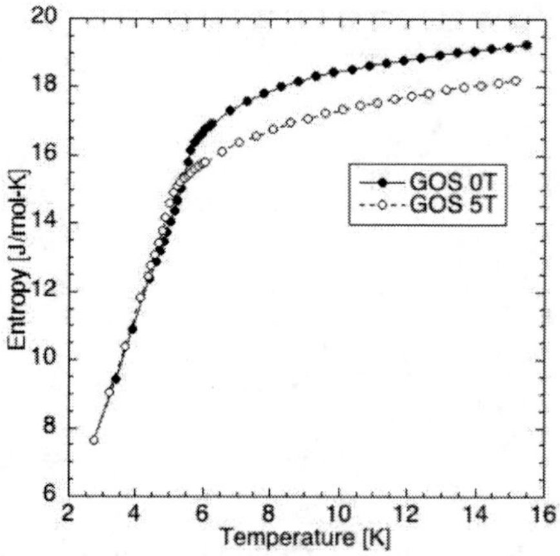

Figure 7. Entropy of Gd2O2S as functions of magnetic field and temperature.

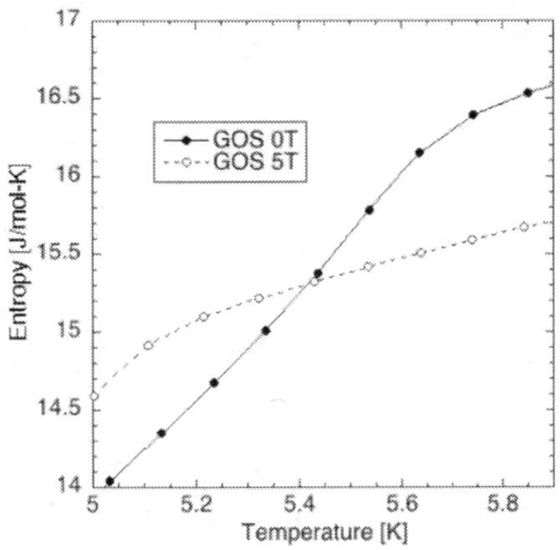

Figure 8. Entropy of Gd2O2S as functions of magnetic field and temperature between 5 and 6 K.

4 Conclusions

We measured the variation of GOS magnetization with external magnetization and temperature. However, we did not observe a spin-flop because and no notable magnetization change occurred when GOS was magnetized to 5 T. Therefore, we assume that a large external magnetic field, for example 10 T is required to observe spin flop behavior.

We measured the variation of the specific heat of GOS with temperature and calculated the entropy change of GOS. Although we could not find an efficient entropy change around 4 K, which is the high temperature stage of an ADR, we confirmed inverse magnetocaloric effects, as shown in Figure 8. Thus, in our future work we will measure the entropy properties of GOS in a high external magnetic field.

References

1. Ishisaki Y et al, "Performance test of Ti/Au bilayer TES microcalorimeter in combination with continuous ADR," AIP Conf. Proc. 1185, 442 (2009).
2. Shirron PJ et al, "A Multi-Stage Continuous-Duty Adiabatic Demagnetization Refrigerator," Adv. Cryo. Eng. 45B, 1629 (2000).
3. Shirron PJ et al, "Proper es of two stage Adiaba c Demagne za on Refrigerator," Cryogenics 41, 789 (2002).
4. Fukuda H et al, "Properties of a two stage adiabatic demagnetization refrigerator", Proceedings of the Cryogenic Engineering Conference (CEC), June 2015.
5. Warburg E. "Magnetische Untersuchungen", Annalen der Physik, 249, 141 (1881).
6. Numazawa T et al, "A New Ceramic Magnetic Regenerator Material for 4 K Cryocoolers," Cryocoolers 12, Kluwer Academic/Plenum Publishers, New York (2002).
7. Numazawa T et al, "Status of the Development of Ceramic Regenerator Materials," Cryocoolers 13, Kluwer Academic/Plenum Publishers, New York (2004).
8. Numazawa T et al, "Magnetocaloric Effect of Polycrystal $GdLiF_4$ for Adiabatic Magnetic Refrigeration", Proceedings of the 24th International Conference on Low Temperature Physics Low Temperature Physics, (2006).
9. Tegus O et al, "Magnetic phase transitions and magnetocaloric effects", Physica B 319, 174 (2002).
10. Zhang Y et al, "Giant magnetoresistance and magnetocaloric effects of the $Mn_{1.82}V_{0.18}Sb$ compound. J. Alloys and Comp. 365, 35 (2004).
11. Krenke T et al, "Inverse magnetocaloric effect in ferromagnetic Ni-MnSn alloys", Nature Materials 4, 450 (2005)

Heat Transfer in a Loop Heat Pipe Using Fe_2NiO_4-H_2O Nanofluid

Prem Gunnasegaran[1,a], Mohd Zulkifly Abdullah[2] and Mohd Zamri Yusoff[1]

[1]*Centre for Fluid Dynamics (CFD), College of Engineering, Universiti Tenaga Nasional, Putrajaya Campus, Jalan IKRAM-UNITEN, 43000 Kajang, Malaysia*
[2]*School of Aerospace Engineering, Universiti Sains Malaysia, Engineering Campus, 14300 Nibong Tebal, Penang, Malaysia*

Abstract. Nanofluids are stable suspensions of nano fibers and particles in fluids. Recent investigations show that thermal behavior of these fluids, such as improved thermal conductivity and convection coefficients are superior to those of pure fluid or fluid suspension containing larger size particles. The use of enhanced thermal properties of nanofluids in a loop heat pipe (LHP) for the cooling of computer microchips is the main aim of this study. Thus, the Fe_2NiO_4-H_2O served as the working fluid with nanoparticle mass concentrations ranged from 0 to 3 % in LHP for the heat input range from 20W to 60W was employed. Experimental apparatus and procedures were designed and implemented for measurements of the surface temperature of LHP. Then, a commercial liquid cooling kit of LHP system similar as used in experimental study was installed in real desktop PC CPU cooling system. The test results of the proposed system indicate that the average decrease of 5.75°C (14%) was achieved in core temperatures of desktop PC CPU charged with Fe_2NiO_4-H_2O as compared with pure water under the same operating conditions. The results from this study should find it's used in many industrial processes in which the knowledge on the heat transfer behavior in nanofluids charged LHP is of uttermost importance.

1 Introduction

As a new kind of heat transfer working fluid, the nanofluid is a new technology attempt to use the special properties of this functional fluid to enhance the phase-change heat transfer in heat pipes, and will have wide application prospect. The term 'nanofluid' refers to a two-phase mixture with its continuous phase being generally a liquid and the dispersed phase constituted of 'nanoparticles' i.e., extremely fine metallic particles of size below 100 nm. As known, water has been widely used as industrial heat transfer fluid for a long time, but due to higher thermal conductivities of metals compared to water, suspensions of nano-sized solid particles especially metal or metallic oxides particles in water, have been used as heat transfer fluid with higher thermal conductivities. Due to this fact, many studies on various water-based nanofluids have been carried out in the past years [1]. For example, application of water-based nanofluids in a heat exchanger exhibits an increased heat transfer coefficient compared to pure water [2, 3]. Seyf and Feizbakhshi [4] study on a numerical investigation of the application of CuO-H_2O nanofluids in micro-pin-fin heat sinks and found a significant

[a] Corresponding author: prem@uniten.edu.my

enhancement in heat transfer. Hajmohammadi et al. [5] consider Cu and Ag water based nanofluids for the flow and heat transfer of nanofluids over a permeable flat plate with convective boundary condition. In the cases of injection and impermeable surface, increasing the nanoparticles volume fraction result in augmentation of convection heat transfer rate. However, in the case of suction, adding Cu and Ag particles reduces the convection heat transfer coefficient at the surface. Overall, many researchers have been devoted to exploiting water-based nanofluids. The research on application of nanofluids in heat pipes was firstly published in 2003 [6]. Many articles have been published since then, involving micro-grooved heat pipe, mesh wick heat pipe, sintered metal wick heat pipe, oscillating heat pipe (OHP), and loop heat pipe (LHP) [7-12]. The applied nano-materials included metal, metal oxides, diamond, carbon nanotubes and several other materials.

The fundamental studies of nanofluids applied in heat pipes are still in its initial stage, most of research works are carried out experimentally to focus on finding out key factors affecting the reliable application of nanofluids in the heat pipes and some experimental results cannot be unified yet. The type, size of heat pipes and operating conditions of heat pipes, the kind of base fluids, the material and size of nanoparticles all varied in very wide ranges among these experiments. Therefore, it is difficult to quantitatively make the comparison among different experimental data and then the most existing research conclusions are qualitative. Study on optimization of operating parameters such as nanoparticle mass concentrations and heat inputs is also rare.

Most of the previous works considered on conventional heat pipes such as micro-grooved heat pipe, mesh wick heat pipe and oscillating heat pipe, and there is far less work conducted for LHPs. As LHPs are demonstrated to be a reliable and great potential for electronic cooling applications, the use of nanofluids could represent a gain in their performance simply by adding a certain volume of nanoparticles to the working fluid. Furthermore, since LHPs utilize the phase change of the working fluid to transport the heat, the selection of nanofluid is of essential importance to promote the thermal performance of LHPs. Due to limited study on thermal performance of LHP charged with nanofluid has been reported in the past, more investigations are needed.

Moreover, heat pipes charged with nanofluids are not only of academic interest but also of industrial interest. Recently, the number of companies that observe the potential of the technology of heat pipes filled with nanofluids is increasing. In the realm of electronics cooling, some companies are conducting research to use nanofluids instead of water for cooling CPUs in desktop computers. Up to now, studies concerning application of nanofluids in LHPs for cooling desktop computers in real life have been limited to academic and theoretical study. Accordingly, in the present study, all the aforementioned issues on application of nanofluid in LHP are addressed experimentally and a real computer setup with a quad-core processor has been used to determine the effect of use of Fe_2NiO_4-H_2O nanofluid on the cooling system in real practical condition.

2 Materials and methods

2.1 The experimental setup

The schematic diagram of experimental setup for LHP under investigation is shown in Fig. 1. The main function of this experiment rig is to determine the thermal performance of LHP charged with Fe_2NiO_4-H_2O nanofluid with mass concentration ranged from 0 to 3% as a working fluid. The LHP shown in Fig. 1 installed with a power supply (W5 Series 30A-720A) and a flat evaporator with a total dimension of 50 mm x 50 mm x 4 mm. A water tank with 0.75 liter glass vessel equipped by drain valve is used as liquid reservoir. The whole LHP is made of copper. The internal and external diameters of both vapor and liquid lines are 13.5mm and 15mm, respectively. The condenser section is made of 50 aluminum rectangular fins and cooled by installing two pieces of screwed fans. To maintain steady state cooling conditions in the condenser section, the temperature and flow rate of the cooling liquid are fixed at constant value. To minimize the heat loss, the whole LHP is insulated by using glass wool. A copper block with heat rods inside is used to simulate the heat source, and the contact area between the evaporator and the heat source is 50 mm x 50 mm. In this experiment, the K-type thermocouples with an accuracy of ± 0.1°C are installed on the pipe/wall in different locations of the loop, including the copper base plate (T_B), the evaporator (T_E), the vapor line (T_V), the condenser section (T_C) and the liquid line (T_L). The temperatures measured by the thermocouples are collected through a data acquisition (Agilent 34970A) with sample rate of 1 Hz and connected to a PC to collect the data. The experiments are conducted under a heat input ranged from 20 W to 60 W by adjusting the variable transformer. The best estimated coolant flow rate and airflow velocity of condenser in present experiment is found to be 5 ml/min and 6 m/s, respectively, are used throughout the entire study. All the sensors are adjusted according to the desired heat input.

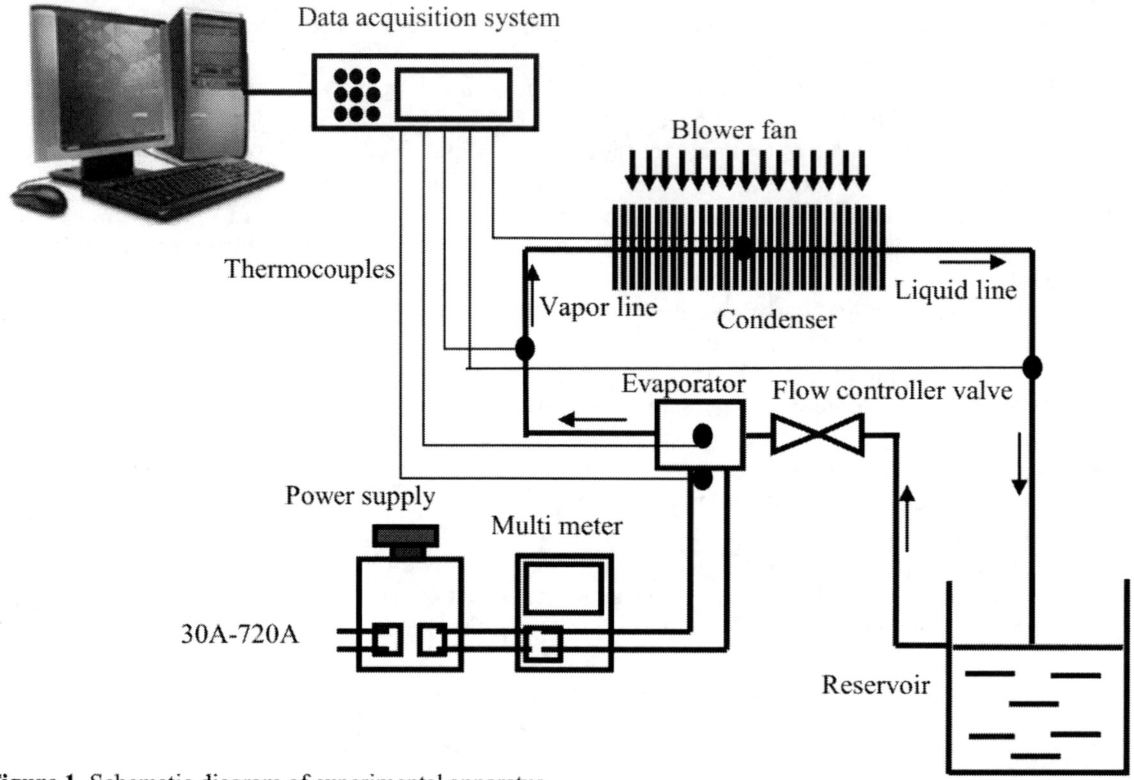

Figure 1. Schematic diagram of experimental apparatus.

2.2 Nanofluid preparation

In the present work, deionized water (DI-water) is taken as the base fluid for preparation of nanofluids in a Digital Ultrasonic Cleaner TJ001. Nanoparticles (Nanopowder) are purchased from Alfa Aesar Corp., Ward Hill, MA and blended with deionized water without any surfactant. The nanoparticles used in the present experiments are Fe_2NiO_4 with an average size of 50 nm and density of 5.37 g/cm^3.

To make a desired percent mass concentration of nanofluids, the weights of deionized water and nanoparticles are measured by a sensitive balance (Ohaus Adventurer Balances) with an accuracy of 0.1 mg. The mass concentration of the powder is calculated from the weight of dry powder using the density provided by the supplier and the total volume of the suspension. For an example, 5.37 g of diamond nanoparticles, which is 1 ml based on the density provided by the vendor, are added to the 99 g (99 ml) of DI-water to make 1 percent mass concentration of Fe_2NiO_4-H_2O nanofluid.

$$\% \text{ mass concentration} = \frac{w_{np}}{w_{bf} + w_{np}} \text{ x } 100\% \qquad (1)$$

Where,

W_{np} = Amount of nanoparticles in gram

W_{bf} = Amount of base fluid in gram.

The nanofluid is then stirred by a magnetic stirrer for 5 hours before undergoing ultrasonication process for one and a half hours. This is to ensure uniform dispersion of nanoparticles and also to prevent the nanoparticles from the aggregation in the nanofluids. Then, the nanoparticle/DI-water mixture is ultrasonicated using an ultrasonic cleaner. The purpose of sonication is to vibrate the nanoparticles in the base fluid so that agglomerates of nanoparticles will break up forming a suspension of isolated nanoparticles.

The nanofluids are at significantly high temperature after the sonication. The nanofluids samples that prepared are kept for observation and no particle settlement is observed at the bottom of the beaker containing nanofluids. Since no concentration gradient appears, the nanofluid employed here maintains stability for several weeks and hence surfactants are not mixed in the prepared nanofluids in present experiment. The nanofluids prepared are assumed to be a homogenous, Newtonian in behavior and their thermophysical properties are uniform and constant with time all through the fluid sample. Fig. 2 shows the transmission electronic microscope (TEM) images of the dispersed nanoparticles in water with particle mass concentration of 0.5% at magnification-105kx at room temperature. It is clear that the Fe_2NiO_4 nanoparticles show the good dispersion and very narrow size distribution.

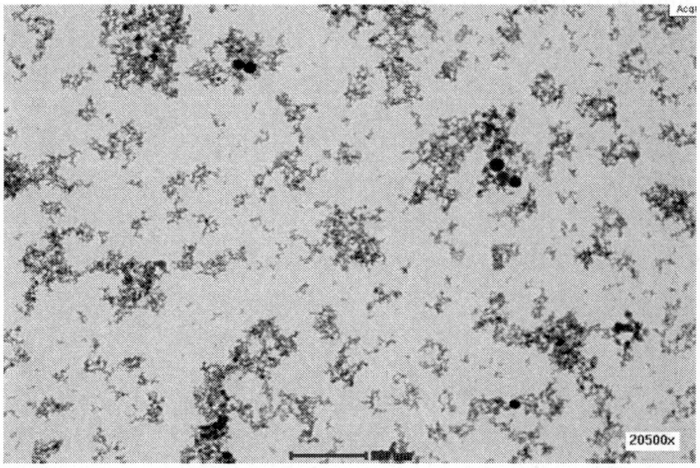

Figure 2. TEM images of Fe_2NiO_4 nanoparticles suspended in pure water.

2.3 Thermal analysis

The objective of the current study is to evaluate the total thermal resistance (R_{th}) and the temperature distributions of the LHP using pure water and Fe_2NiO_4-H_2O nanofluid as working fluids for various heat inputs under steady state and transient conditions. The thermal resistance network of the system is shown in Fig. 3.

The heat flux ($\dot{q}$) that applied on the bottom of base plate can be expressed as:

$$\dot{q} = \frac{Q}{A_b} \quad (2)$$

where Q denotes the heat input and A_b is the area of base plate. The thermal resistances of the LHP are defined by Franchi and Huang [13]:

The thermal resistance between the copper base plate and the evaporator section (R_B) is:

$$R_B = \frac{T_B - T_E}{Q} \tag{3}$$

where T_b denotes the temperature at the copper base plate and T_E is the temperature at the evaporator.

The thermal resistance of the evaporator section (R_E) is:

$$R_E = \frac{T_E - T_V}{Q} \tag{4}$$

where T_V is the temperature at the vapor line.

The thermal resistance of the vapor line (R_V) is:

$$R_V = \frac{T_V - T_C}{Q} \tag{5}$$

where T_C is the temperature at the condenser section.

The convective thermal resistance of the condenser (R_C) is:

$$R_C = \frac{T_C - T_L}{Q} \tag{6}$$

where T_L is the temperature at the liquid line.

The thermal resistance of the liquid line (R_L) is:

$$R_L = \frac{T_L - T_A}{Q} \tag{7}$$

where T_A is the ambient temperature.

According to the thermal resistance network as shown in Fig. 3, the R_{th} of the system is given by:

$$R_{th} = R_B + R_E + R_V + R_C + R_L \tag{8}$$

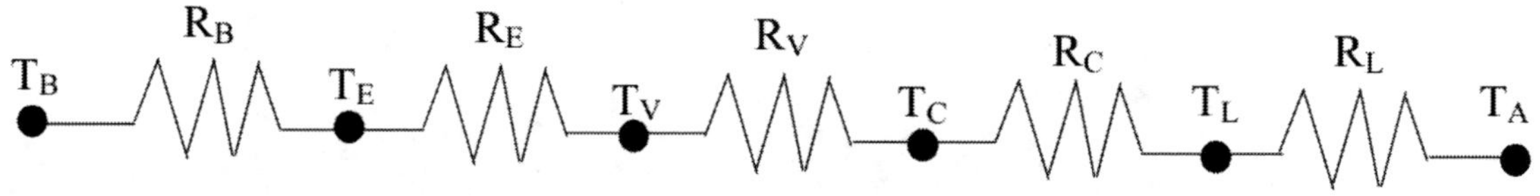

Figure 3. Thermal resistance network of LHP.

3 Proposed desktop PC CPU cooling system

The current work investigated the cooling performance of a water cooling kit with the addition of nanofluids in desktop PC CPU system. A real computer setup with a quad-core processor has been used to determine the effect of use of nanofluid on the cooling system in real practical condition. For the system setup, commercial liquid cooling kit of loop heat pipe (LHP) system similar as used in experimental study was installed in CPU. The system consisted of a quad core processor (Intel Core i5 3470). In the computer, a HeavyLoad software is used to stress all resources of a PC (like CPU, GPU, RAM, hard disk, operating system, etc.) to the maximum from normal condition in order to test if it will run reliably under heavy load (100% CPU usage). To stress the PC or server, HeavyLoad writes a large test-file to the temperature folder, it allocates physical and virtual memory, performs complex calculations and it draws patterns in its window. The temperature of the processor is measured by the internal temperature sensor while applying a constant processing load (maximum load = 100%) on the processor resulting in the highest operating temperature. The highest operating temperature of the processors in nanofluid cooled-CPU was recorded and then compared with conventional water cooled-CPU.

4 Results and discussion

4.1 Thermal resistance analysis

Fig. 4 shows the relationship between the R_{th} and nanoparticle mass concentration for various heat input. As depicted in Fig. 4, the R_{th} decreased up to 1% and increased on further increase in nanoparticle mass concentration (from 1% to 3%) at all heat inputs. Thus, there is an optimal particle concentration, which is about 1% for the Fe_2NiO_4-H_2O-charged LHP in the present experiment. Hence, the addition of Fe_2NiO_4 nanoparticles to base water more than 1% is not recommended for future study. At the optimal mass concentration of 1%, the maximum reduction in R_{th} of Fe_2NiO_4-H_2O charged LHP is about 2.654 °C/W (or 7.17%) under a heat input of 60W, is obtained as compared with pure water (0% nanoparticle mass concentration) charged LHP. It is clear from the Fig. 4 that when the heat input increases, the thermal resistance decreases and the nanoparticle mass concentration has a great impact on the R_{th}.

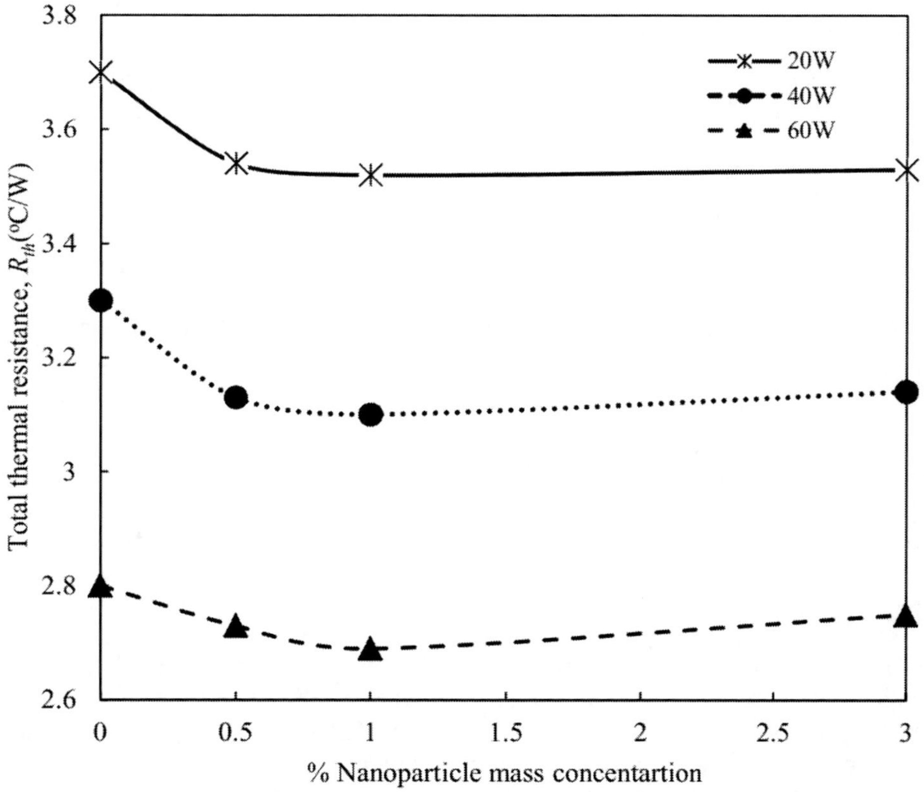

Figure 4. Influence of Fe_2NiO_4 nanoparticle mass concentrations on the R_{th} of LHP for various heat inputs.

4.2 Processor operating temperature of desktop PC CPU system

Decreasing the processor operating temperature (core temperature) of desktop PC CPU is the main purpose of using nanofluids in current proposed liquid cooling systems. This temperature is measured while applying a maximum constant processing load on the processor by using HeavyLoad software

DOI: 10.1051/matecconf/201710905001

resulting in the highest operating temperature. HeavyLoad has been designed to test a computer's resistance to extreme operational conditions, such as intense CPU and RAM usage to avoid server crashes or to make sure the computer will not over heat. The internal temperature sensor of the processor is used to measure the temperature. In the current proposed systems, the pure water and Fe_2NiO_4-H_2O nanofluid is selected as a working fluid within the LHP in CPU. The Fe_2NiO_4-H_2O nanofluid with nanoparticle mass concentration of 1% is selected as the experimental results show that this type of nanofluid yields the best thermal performance. The measured core temperatures are illustrated in Table 1. As expected, adding of the Fe_2NiO_4-H_2O nanofluid with particle mass concentration of 1% to the cooling system reduced the core temperature in comparison to when pure base fluid (water) is used. The core temperature difference between pure water and Fe_2NiO_4-H_2O nanofluid and percent reduction are illustrated in Table 2. The test results showed the average decrease of 5.75°C (14%) in core temperature of desktop PC CPU with Fe_2NiO_4-H_2O nanofluid cooling system as compared with pure water.

Table 1. Processor operating temperatures.

CPU: Processor 4 Core(s)
Processor Information

Model Intel Core i5 3470 (Ivy Bridge)

Processor #	Temperature Readings for LHP charged with pure water	Temperature Readings for LHP charged with Fe_2NiO_4-H_2O nanofluid	Load
Core #0	43 °C	37 °C	100%
Core #1	43 °C	38 °C	100%
Core #2	42 °C	37 °C	100%
Core #3	39 °C	32 °C	100%

Table 2. Percent reduction in processor operating temperatures of using Fe_2NiO_4-H_2O nanofluid.

Processor #	Temperature difference between pure water and Fe_2NiO_4-H_2O nanofluid	% Reduction using Fe_2NiO_4-H_2O nanofluid
Core #0	6 °C	14.0
Core #1	5 °C	11.6
Core #2	5 °C	11.6
Core #3	7 °C	18.0

5 Conclusions

Experimental investigation on Fe_2NiO_4-H_2O nanofluid flow and heat transfer characteristics in LHP are reported in this study. The effects of using nanofluid with different particle mass concentrations at different heat inputs on the heat transfer enhancement of LHP are comprehensively studied. Based on present experimental findings, the most appropriate type of nanofluid to be utilized for the cooling system in real practical condition by using a real computer setup is Fe_2NiO_4-H_2O nanofluid with particle mass concentration of 1%. Fe_2NiO_4-H_2O nanofluid with particle mass concentration of 1% is preferable and recommended as a superior heat transfer fluid in a LHP for desktop PC CPUs cooling system as it has the lowest R_{th} among the other nanoparticle mass concentrations studied. The proposed desktop PC cooling system for industrial applications in current study, results show enhanced heat transfer in the cooling of computer microchips, as indicated in reduction of the operating temperature of processor (Core-i5 processor) when using the Fe_2NiO_4-H_2O nanofluid as compared to application of pure water under 100% of CPU usage. This finding suggests that no extra cooling fans or modifications on geometrical parts of LHP is required to dissipate the excess heat as nanofluid itself is sufficient enough to keep the computer from overheating. This would avoid noise generation by the cooling system and reduce the operation and maintenance costs as well. Therefore, nanofluid should be considered to satisfy the economy of the computer cooling system for real practical operating conditions.

Acknowledgments

The authors would like to sincerely thank the Ministry of Higher Education (MOHE) of Malaysia for the provision of a grant with code no 20160103FRGS to support this work.

References

1. L. Fedele, L. Colla, S. Bobbo, S. Barison, F. Agresti, Experimental stability analysis of different water-based nanofluids, Nanoscale Res. Lett., **6**, 300–308 (2011).
2. G. Huminic, A. Huminic, Application of nanofluids in heat exchangers: are view, Renew. Sust. Energy Rev., **16**, 5625–5638 (2012).
3. S. Halelfadl, A. M. Adham, N. Mohd Ghazali, T. Mare P. Estelle R. Ahmad, Optimization of thermal performances and pressure drop of rectangular microchannel heat sink using aqueous carbon nanotubes based nanofluid, Appl. Therm. Eng., **62**, 492–499 (2014).
4. H. R. Seyf, M. Feizbakhshi, Computational analysis of nanofluid effects on convective heat transfer enhancement of micro-pin-fin heat sinks, Int. J. Therm. Sci., **58**, 168–179 (2012).
5. M. R. Hajmohammadi, H. Maleki, G. Lorenzini, S. S. Nourazar, Effects of Cu and Ag nano-particles on flow and heat transfer from permeable surfaces, Adv. Powder Technol., **26**, 193–199 (2015).
6. Chien, H. T., Tsai, C. Y., Chen, P. H. and Chen, P. Y., Improvement on thermal performance of a disk-shaped miniature heat pipe with nanofluid, *Proceedings of the Fifth International Conference on Electronic Packaging Technology, IEEE*, Shanghai, China, 389-391 (2003).
7. G. S. Wang, B. Song, Z. H. Liu, Operation characteristics of cylindrical miniature grooved heat pipe using aqueous CuO nanofluids, Expt. Therm. Fluid Sci., **34**, 1415–1421 (2010).
8. A. B. Solomon, K. Ramachandran and B. C. Pillai, Thermal performance of a heat pipe with nanoparticles coated wick, Appl. Therm. Eng., **36**, 106-112 (2012).
9. M. Shafahi, V. Bianco, K. Vafai, and O. Manca, An investigation of the thermal performance of cylindrical heat pipes using nanofluids, Int. J. Heat. Mass. Transf., **53**, 376-383 (2010).
10. L. G. Asirvatham, R. Nimmagadda, S. Wongwises, Heat transfer performance of screen mesh wick heat pipes using silver-water nanofluid, Int. J. Heat. Mass. Transf., **60**, 201–209 (2013).
11. H. B. Ma, C. Wilson, Q. Yu, K. Park, U. S. Choi, M. Tirumala, An experimental investigation of heat transport capability in a nanofluids oscillating heat pipe, J. Heat. Transf., **128**, 1213–1216 (2006).
12. Z. Wan, J. Deng, B. Li, Y. Xu, X. Wang and Y. Tang, Thermal performance of a miniature loop heat pipe using water-copper nanofluid., Appl. Therm. Eng., **78**, 712-719 (2015).
13. G. Franchi and X. Huang, Development of composite wicks for heat pipe performance enhancement, Heat Transf. Eng., **29**, 873–884 (2008).

Study on Experiment of Concrete Strength with Numerical Simulation

Xiao Sun[1, 2,a]

[1]*Nanchang Institute of Technology, Hydraulic and Ecology Engineering Department, 330099 Nanchang, China*
[2]*Jiangxi Engineering Research Center of Water Engineering Safety and Resources Efficient Utilization, 330099 Nanchang, China*

Abstract. The limit state of bearing capacity of concrete and collapse state is discussed by the overload method of finite element limit analysis and finite difference software FLAC3D. Firstly, the shear strength of concrete with different strength grade is verified through the numerical limit analysis. Besides, a new method to judge the failure of concrete by calculating the concrete limit shear strain increment value is proposed. In addition, the collapse surface position and shape is simulated and the result is compared with the actual state of the destruction. The finite element limits analysis method have been tried to applied in concrete in this paper.

1 Introduction

The Concrete is the largest and most widely used engineering materials, which is applied in many fields such as construction, hydraulic, transportation, national defense and so on. In the architectural mechanics, concrete has properties of tensile failure, compression failure, flexural failure, but in mechanics, the destruction modes of the solid materials only have tensile failure and shear failure[1-2]. Currently, in some projects, some concrete is mainly in plastic state, such as petroleum engineering, hydraulic engineering, tunnel engineering, its bearing capacity and failure state need to be analyzed based on the elastic-plastic theory. According to the view of mechanics, concrete belongs to the geotechnical friction material, which is not only has cohesion but also friction, whose shear strength needs to be pressed by cohesion c and frictionφ. It is very important to control the strength index of concrete. Considering that, there is no unified specification and measurement methods for c, φ of concrete and that, the cohesion and internal friction angle of soil and rock is mainly measured by direct shear test and triaxial test, so we obtain the cohesive and internal friction angle of concrete with the help of the shear test and triaxial test following the example of geotechnical material [3-6]. Due to the lack of corresponding test instrument, we have proposed to determine c, φ of concrete by combining direct shear test and uniaxial compression test.

The purpose of this article is to judge the characteristics of the ultimate bearing capacity and damage state by performing numerical simulations of concrete strength test. The overload method of finite element limit analysis and the finite difference software FLAC3D is applied in the numerical simulation. For all this, the shear strength of concrete specimens with

[a] Corresponding author: xiaosun@nit.edu.cn

DOI: 10.1051/matecconf/201710905002

different strength grade is verified first. Then extract the ultimate shear strain increment of concrete with different strength grade and identify whether the concrete is damaged based on the limit value, providing new failure criterion for finite element limit analysis method. At last, the position and form of concrete specimen failure surface is simulated, and compare the result with the actual damage state.

In this paper, we cannot only get ultimate load and safety factor but also the position and shape of the structure failure surface by application of finite element limit analysis method. It has important academic significance and application value for the study of the force characteristics of concrete structure.

2 Numerical simulation and experiment of concrete shear strength

2.1 The calculation model and parameters

The In order to verify that compressive strength is unrelated to geometric size and shape of the model in the test above, we perform numerical simulations of 3 specimens with different size and shape based on the overload method of finite element limit analysis and finite difference software FLAC3D.

Scheme 1: model is established according to the concrete cube specimen from "specification for design of concrete structures"(GB50010-2011). The size is 150 mm ×150 mm × 150 mm. Enforce vertical constraints on the bottom, and enforce vertical downward uniform distributed load on the surface, as shown in figure 1. Concrete is regarded as the ideal elastic-plastic material. The ideal elastic-plastic constitutive model and Moore - coulomb yield criterion is adopted in the calculation. Four concrete models with C_{25}, C30, C35, C40 grade are calculated. The c and φ of concrete is obtained from the triaxial test. Extract the elastic modulus according to "specification for design of concrete structures"(GB50010-2011), as shown in table 1. The calculation model is shown in figure 2.The similarity ratio between calculation model and test model is 1:1.

Table 1. Physico-mechanical Parameters of concrete.

Surrounding rock type	Modulus of elasticity E/GPa	Poisson's ratio v/1	Gravity ρ/kN.m-3	Cohesion C/MPa	Fraction angle φ /°
C_{25}	28.0	0.2	2500	3.2	61.3
C_{30}	30.0	0.2	2500	3.9	61.8
C_{35}	31.5	0.2	2500	4.5	62.2
C_{40}	32.5	0.2	2500	5.1	62.5

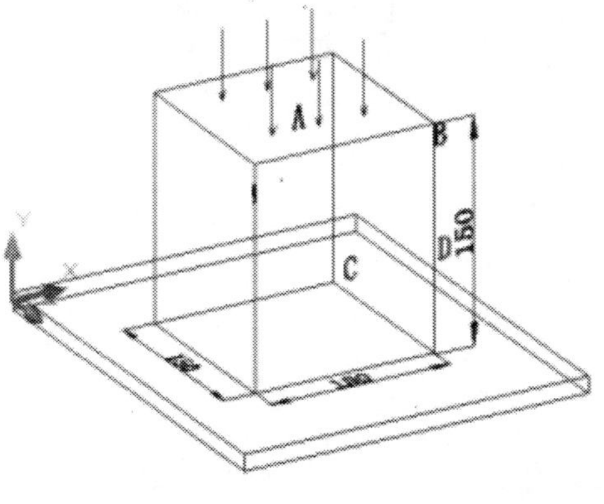

Figure 1. Geometric model.

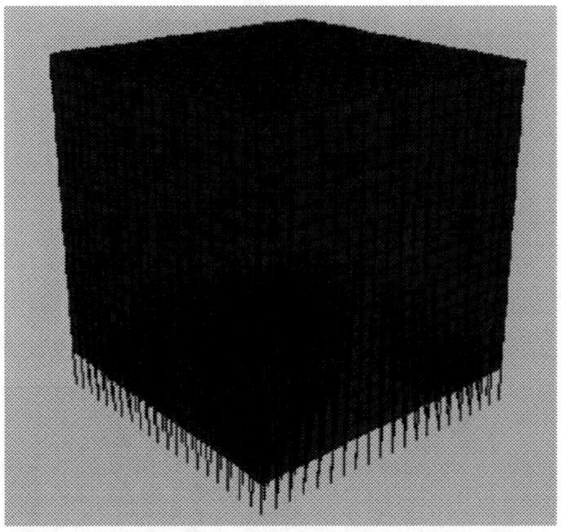

Figure 2. Calculation model.

2.2 Comparative analysis of the measured result and the calculated result

Modify the load constantly during the calculation. The load when the model reaches the limit damage state is defined as the calculated compressive strength of concrete. Determine whether the model is damaged during the calculation according to the three criteria of rock and soil damage that whether the plastic zone runs through, whether the calculation converges and whether the displacement or strain mutates. During the calculation, the displacement of point A (0.075, 0.075, 0.15) in Z direction, the displacement of point B (0.15, 0.00, 0.15) in Y and Z direction, the placement of point C (0.075, 0.00, 0.075) in Y direction and the placement of point C (0.075, 0.00, 0.075) in Y direction is monitored. Figure 3 shows the displacement-time curve when the calculation converge, the curve is horizontal in the end. It indicates that the displacement change won't happen again and the calculation converges. Figure 4 shows the displacement-time curve when the calculation doesn't converge, it can be seen that the displacement mutate when iteration is about 2000 times, displacement shows a trend of increase, calculation won't converge. It indicates that C_{25} concrete specimen has been destroyed.

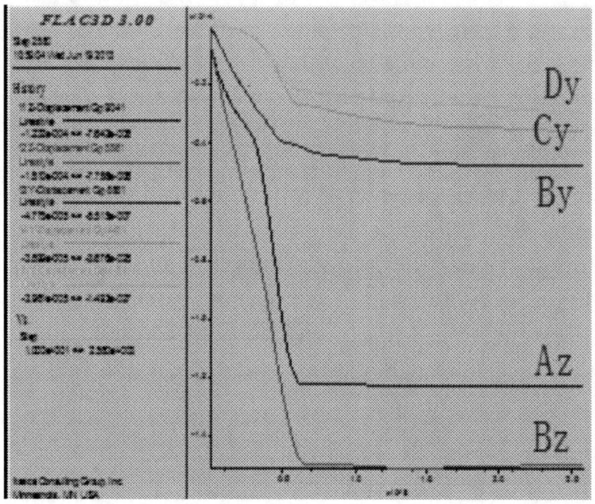

Figure 3. The displacement-time curve of key points when calculation converges the displacement-time

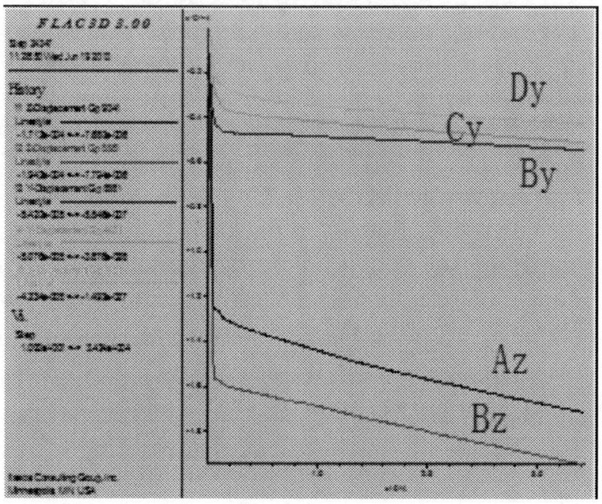

Figure 4. Curve of key points when calculation converges doesn't converge.

The comparison of measured values and calculated values of compressive strength in Scheme Iis shown in table 2.

Table 2. The comparison of measured values and calculated values of compressive strength in Scheme I.

The concrete strength grade	The calculated value of compressive strength (MPa)	The measured value of compressive strength (MPa)	The error between calculated value and measured value (%)
C_{25}	25.01	25.00	0.04
C_{30}	31.05	30.00	3.50
C_{35}	36.36	35.00	3.89
C_{40}	41.68	40.00	4.20

Specimens in scheme 2 are cubes with the size of 1500 mm×1500 mm×1500 mm, other conditions are the same as scheme 1. The calculated result is shown in table 3.

Table 3. The comparison of measured values and calculated values of compressive strength in Scheme II.

The concrete strength grade	The calculated value of compressive strength (MPa)	The measured value of compressive strength (MPa)	The error between calculated value and measured value (%)
C_{25}	24.98	25.00	0.08
C_{30}	31.02	30.00	3.40
C_{35}	36.33	35.00	3.80
C_{40}	41.66	40.00	4.15

It can be seen through the calculation above that the measured values and the calculated results of compressive strength are particularly close. The maximum error between the two is 4.20%, the minimum error is 0.04%. There is a small difference between the compressive strength measured values and the calculated values in different schemes with the same strength grade, it indicates that compressive strength is unrelated to the size and shape of material. It also shows that, the concrete strength parameters c and φ measured by the test are reasonable and the finite element limit analysis method which is widely applied to geotechnical engineering is also appropriate for concrete material.

2. 3 The theoretical relationship between the shear strength and compressive strength of concrete and its verification

Due to the inconsistency development of the mechanics, in architectural mechanics which is applied to bar and artefacts, the strength of materials is usually determined by the force form of component, for example, compressive strength, tensile strength, shear strength and so on. But it is usually determined by the form of material damage in the elastic and plastic mechanics. The shear failure results from pressing in mechanics, so the material strength parameter adopts tensile strength and shear strength, it doesn't adopt compressive strength. In fact, the two are the same essentially, they just have different definition. But the shear strength of friction materials is composed of cohesion and friction. There must exist corresponding mechanical relationship between shear strength and compressive strength. For the friction materials, the following relationship can be deduced out on the basis of Mohr-Coulomb criterion:

$$\sigma_1 = \frac{1+\sin\varphi}{1-\sin\varphi}\sigma_3 + \frac{\cos\varphi}{1-\sin\varphi}2c \tag{1}$$

$$\tau = c + \sigma\tan\varphi = \frac{\sigma_1-\sigma_3}{2}\cos\varphi \tag{2}$$

$$\sigma = \frac{\sigma_1+\sigma_3}{2} - \frac{\sigma_1-\sigma_3}{2}\sin\varphi \tag{3}$$

$$c = \left[\frac{\sigma_1-\sigma_3}{2} - \frac{\sigma_1+\sigma_3}{2}\sin\varphi\right]\frac{1}{\cos\varphi} \tag{4}$$

Where: σ —Normal stress on the shear plane.

τ — The limit of shear strength.

c, φ — Cohesive and angle of internal friction.

σ_1 , σ_3 —The maximum and the minimum principal stress.

Concrete compressive strength test is uniaxial test, namely, σ_3 =0,it can be obtained according to (3), (4):

$$\frac{\sigma \tan \varphi}{c} = \sin \varphi \tag{5}$$

Thus, the value of c can be obtained by direct shear test without normal pressure, then work out the value of φ . If the shear strength and compressive strength of concrete are known, of the strength can also be verified according to the above formula. Therefore, the accuracy of c and φ of concrete with different strength grade can be verified according to the above theoretical formula, as shown in table 4. The value compressive strength and shear strength with theoretical solution and numerical solution is shown in table 4, the contribution of cohesion and friction is listed in table 8. It can be seen from table 5 and table 6, that friction accounts for 47 percent and cohesive accounts for 53 percent. The accuracy of the results in table 4 can be proved by formula (5). It can be seen, concrete compressive strength takes rubbing strength into consideration.

Table4. The verification of shear strength of different strength concrete.

Strength grade of concrete（The test results）	c	φ	Ultimate load		Error between theoretical solution and numerical solution (%)
			Theoretical solution σ_1	Numerical solution σ_1	
C_{20}	2.6	60.09	20.03	20.03	-0.0238
C_{25}	3.2	61.3	25.02	25.01	0.0278
C_{30}	3.9	61.8	31.05	31.05	0.0100
C_{35}	4.5	62.2	36.37	36.36	0.0201
C_{40}	5.1	62.5	41.68	41.68	0.0095

Table 5. The value compressive strength and shear strength with theoretical solution and numerical solution

Strength grade of concrete（the test results）	compressive strength σ_1		Shear strength τ	
	Theoretical solution	Numerical solution	Theoretical solution	Numerical solution
C_{20}	20.03	20.03	4.87	4.87
C_{25}	25.02	25.01	6.01	6.01
C_{30}	31.05	31.05	7.34	7.34
C_{35}	36.37	36.36	8.48	8.48
C_{40}	41.68	41.68	9.62	9.62

Table 6. The value cohesion and friction contributes to the shear strength of different strength concrete.

Strength grade of concrete（the test results）	Cohesion c			Friction $\sigma \tan \varphi$		
	Theoretical solution	Numerical solution	Rate	Theoretical solution	Numerical solution	Rate
C_{20}	2.6	2.6	0.53	2.27	2.27	0.47
C_{25}	3.2	3.2	0.53	2.81	2.81	0.47

C_{30}	3.9	3.9	0.53	3.44	3.44	0.47
C_{35}	4.5	4.5	0.53	3.98	3.98	0.47
C_{40}	5.1	5.1	0.53	4.52	4.52	0.47

3 The analysis of limit shear strain increment

The loading process of the solid materials goes through three stages, elastic, plastic and destroyed. The material will spring back to its normal shape in the elastic stage, but it won't recover in plastic stage. It can be judged from the yield criterion whether the material is elastic or plastic. The Tresca criterion or Mises criterion is applied to metal material and Mohr-Coulomb criterion is applied to geotechnical friction material. The yield state does not equal to failure state. It also can bearing load. But it will enter destroyed stage when the plasticity develops finitely. There is no clear criterion to judge whether the material is overall damaged. But we can judge whether the material is destroyed according to the global failure criterion in the limit analysis of material. As shown in figure 5, if the ideal plastic stress-strain curve is expressed by stress, it manifests as the infinite increasing of strain and the remaining constant of stress. The yield criterion and failure criteria is consistent at this point, it is difficult to distinguish damage. If the curve is expressed by strain, it is initial yielding at the beginning of yielding, the materials possess elastic limit strain εy, it will not fail. But when the strain reaches to failure, it possess plastic limit strain εf, the material will fail. It shows that, the limit value of shear strain exists during the failure of concrete. When the shear strain of each point on the failure surface is greater than the limit shear strain, material is known as overall failure, so it can be seen as the criterion for determining the overall failure of material. The limit shear strain is related to the strength and deformation characteristics of material, the property of different strength concrete is consistent, and then we can get the corresponding limit shear strain. And, the variation of properties of different strength concrete is minimal. Their limit shear strain is similar. It is should be indicated that, the limit shear strain of specimen will increase under the action of confining pressure due to the existing concrete fiction. But the confining pressure around the general structures does not exist or it is tiny, the change of the limit shear strain can be ignored.

The paper will give the limit strain of different strength concrete through the finite element limit analysis. Shear strain is presented by the increment in FLAC. The shear strain increment is similar to the shear strain in the mall deformation, so the shear strain increment is the cumulative shear strain increment, not the increment of the current step. This paper utilizes the software FLAC, so the limit shear strain given means the cumulative shear strain increment.

In the above calculation of concrete specimens with three kinds of size and shape, the cohesion c , internal friction Angle φ , bulk modulus K and shear modulus G of different strength concrete differs from each other. The calculation made the model to the limit equilibrium state by overload of the finite element analysis method, now we can get limit load and limit shear strain increment of the model. The shear strain increment will reach the value that can make the model destroyed by increasing the load slightly and repeating the calculation. The shear strain increment when the model failure is greater than the value of the limit equilibrium state. The results are shown in table 7, table 8.

Table 7. The shear strain increment contrast of Scheme I.

. The concrete strength grade	The calculated compressive strength (MPa)	The minimum range of shear strain increment (‰)	
		Limit state	Damaged
C_{25}	25.01	0.980----1.000	1.063----1.200
C_{30}	31.05	1.058----1.100	1.269----1.613
C_{35}	36.36	1.306----1.350	1.363----1.400
C_{40}	41.68	1.419----1.500	1.525---1.513

Table 8. The shear strain increment contrast of Scheme II.

The concrete strength grade	The calculated compressive strength (MPa)	The minimum range of shear strain increment (‰)	
		Limit state	Damaged
C_{25}	24.98	0.972----1.000	1.066----1.100
C_{30}	31.02	1.208----1.250	1.301---1.350
C_{35}	36.33	1.284---1.300	1.381---1.400
C_{40}	41.66	1.454----1.600	1.614---1.620

The result shows that, the limit shear strain increment of concrete with the same strength grade in three schemes is similar to each other. As shown in table9,The value in scheme 1 is 1.000‰—1.063‰, the value in scheme 2 is1.000‰—1.066‰. The maximum range is 0.950‰—1.109‰, it can be regarded as the limit shear strain increment of C25 concrete. Similarly, the limit shear strain increment can be determined as 1.100‰—1.301‰ for C30, 1.300‰—1.426‰ for C35 and 1.500‰—1.614‰ for C40. The limit shear strain increment of different strength concrete will increase along with the increasing of strength, but it change slightly.

Table 9. Setting Word's margins.

value range / scheme	C25	C30	C35	C40
Scheme 1	1.000—1.063	1.100—1.269	1.350—1.363	1.500—1.525
Scheme 2	1.000—1.066	1.250—1.301	1.300—1.381	1.600—1.614
The range of limit shear strain increment	0.950—1.109	1.100—1.301	1.300—1.426	1.500—1.614

4 The failure surface comparison of test and numerical simulation

The failure surface shape of C_{25} grade concrete specimen is shown in figure 6, the specimen appears as X truncated cone after the destruction. In order to find out the accurate failure surface position of concrete, first calculate the range of the C_{25} concrete limit shear strain increment, which is 0.950—1.109‰, so 1.000‰ can be regarded as the value of the limit shear strain increment. Then judge the destroyed cell in the model according that. This process can be realized in the post-processing by programming with FLAC's native FISH language.

Clearly, intercept half of the model by vertical plane. It is the shear strain increment cloud after the above processing, as shown in figure 6. The shear strain increment on the failure surface ranges from 1.000 ‰ to 1.500 ‰. The failure surface appears to be X from the figure. The shear strain increment shows a trend of increase from the failure surface to the surface of the specimen, and it indicates that the area outside the failure surface is entirely destroyed. The shear strain increment shows a trend of decrease from the failure surface to the inside of the specimen and it indicates that the area inside the failure surface is not destroyed. The deepest edge of the destroyed part in the numerical simulation is 10.19 cm from the upper surface and 4.67cm from the right surface. In the test, it is 10.74cm from the upper surface and 3.93cm from the right surface. The form and damage range of the failure surface of the calculation is similar to the result of the test. It shows that, the form and position of the failure surface can be obtained by the application of the finite element limit analysis method.

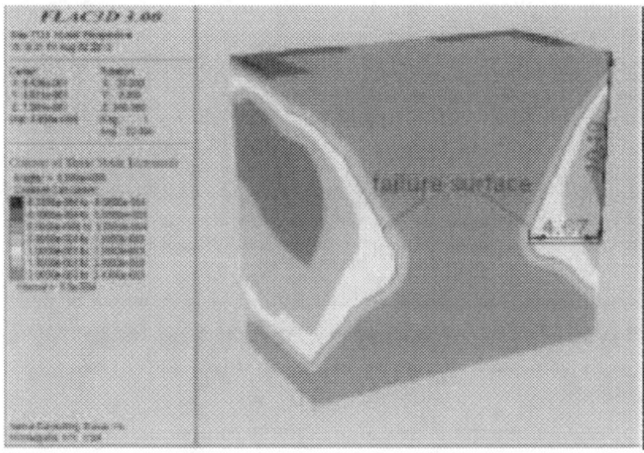

Figure 5.Failure surface of numerical simulation.

Figure 6. Failure surface of concrete test.

5 Conclusions

This article is attempted to obtain the ultimate load and safety factor of the concrete material by the application of finite element limit analysis method. Hope to be able to cause the attention of academia. The conclusions were obtained as follow:

(1) The calculated value of the concrete comprehensive strength is consistent with the measured value, the maximum error is 4.20% and the minimum error is 0.04%. It shows that the concrete strength parameters c and φ determined by the test is reasonable.

(2) The shear strain increment of different strength concrete in the limit failure state is calculated. The interval value of the limit shear strain increment is given. It can be regarded as the criterion for the overall damage of the concrete. Provide a new criterion to determine whether the concrete is broken.

(3) The form and damage range of the failure surface of the calculation is similar to the result of the test. It shows that, the failure surface can be obtained by the application of the finite element limit analysis method.

Acknowledgement

This work was supported by grant of No.GJJ151111 from Youth Project of the Education Department of Jiangxi Province in China.

References

1. ZHENG Ying-ren, Kong Liang,Geotechnical Plastic Mechanics, Beijing: China Building industry Press, (2010)
2. ZHENG Ying-ren, ZHU He-hua, FANG Zheng-chang, The stability analysis and design theory of surrounding rock of underground engineering, Beijing: China Communications Press, (2012)
3. National Standard of the People' s Republic of China, Standard for evaluation of concrete compressive strength(GB/T50107-2010),Beijing: China Architecture & Building Press, (2010)
4. National Standard of the People' s Republic of China,Standard for test method of mechanical properties on ordinary concrete(GB/T50081-2002),Beijing: China Architecture & Building Press, (2003)
5. National Standard of the People' s Republic of China,Code for design of concrete structures(GB/T50010-2010),Beijing: China Architecture & Building Press, (2011)
6. ZHANG Qi, GUO Zhen-hai. Research of shear strength and shear deformation of concrete, Journal of Building Structures, 13(5),pp: 17 – 24,(1992)

Investigation on Coal Slagging Characteristics and Combustion Behaviour in Furnace

Nor Jauhara Sophia[1] and Hasril Hasini[1,a]

[1] *Fluid Dynamics and Risk Reduction Research Group, Institute of Sustainable Engineering, Universiti Tenaga Nasional, Jalan IKRAM-UNITEN, 43000 Kajang, Malaysia*

Abstract. This paper describes an investigation of coal characteristics and slagging potential using coal slagging assessment tool such as numerical slagging indices. Historical data on coal use according to type and shipment in power plant was use for the analysis. It is found that higher content of silica oxide results in higher softening temperature and the monomers modifiers can alter the softening temperature. Calculation also reveals encouraging conditions that could possibly prevent the slagging potential inside the furnace.

1 Introduction

Slagging in the furnace poses problem to coal power station at it reduce the efficiency of heat transfer to water in water tubes. The right viscosity of coal at high temperature has to be in the correct range to ensure good combustion performance of the coals. The ash analysis coals slagging numerical indices has been studied, tested and integrated with actual plant data to obtain numerical slagging indices [1-6]. The importance of the slagging numerical indices is to segregate and classify the coals based on its ash analysis and predict its tendency to change physical properties to slag. In this work, attempt was made to analyse and rank the coals of several type, brand and rank of coals. Coals are ranked to sub-bituminous or bituminous based on its carbon content. This study focused on the types of coals commonly used in one of the biggest power plant in Malaysia. Recent operational problem has shown that large chunk of coal slag was formed and subsequently fall to the bottom hopper of the boiler, causing significant damage to the furnace. Hence, analysis of the formation of slagging using historical operational data is carried out to establish the understanding in terms of ash analysis and tendency to slag, by taking into account the temperature correlation between the viscosity, mineral components and either guiding parameters.

2 Materials and methods

2.1 Numerical Slagging Indices

The coals quality database was obtained from plant historical operation data. The database includes crucial data for coal certificate of ultimate analysis and proximate ash analysis. The useful data were

[a] Corresponding author: hasril@uniten.edu.my

Table 1 Summary of some existing slagging indices [1-6]

Author (s)	Basis	Slagging Indices	Slagging potential			
			low	Medium	High	Severe
Lawrence et al. 2008	Silica Ratio	$SR = 100(SiO_2/SiO_2 + Fe_2O_3 + CaO + MgO$	72 – 80	65 – 72	50 – 65	-
Lawrence et al. 2008	B/A Ratio	$B/A = \frac{Fe_2O_3 + CaO + MgO + K_2O - Na_2O}{SiO_2 + Al_2O_3 + TiO_2}$	< 0.11	0.11 – 0.14	> 0.14	-
Lawrence et al. 2008	Sulphur Content	$Index = (B/A) \times S$	< 0.6	0.6 – 2.0	2.0 – 2.6	> 2.6
Lawrence et al. 2008	Shrinkage Rate	$Index = \frac{Rm}{Lm} \times Wt$	No standard scale shown, but from the report ≤ 55 shows no slagging relative to others			
McLennan et al. 2000	Fusibility	$AF = \frac{4IDT + HT}{5}$	> 1343°C	1232 – 1343°C	1149 – 1232°C	< 1149°C
McLennan et al. 2000	Percentage of Fe_2O_3	$Index = \%Fe_2O_3\ in\ ash$	< 6%	6 – 7%	> 7%	-
Watt and Fereday 1969	Viscosity	$Log\,(\mu) = \frac{m \times 10^7}{(T-150)^2} + c$	High particle viscosity yields low slagging potential			
Yin et al. 1998	Fusibility	$ST = a(SiO_2) + b(Al_2O_3) + c(Fe_2O_3) + d(CaO) + e(MgO) + f(x) + g$	High fusibility yields low slagging potential			

used to conduct calculation of empirical numerical slagging indices and the combustion performance studies using computational fluid dynamics at a later stage.

Slagging indices has been studied previously by other researchers to find the correlations to properly predict the collision behaviour of the ash. This was done by understanding its fusibility and viscosity of coal ash in the furnace and the ash particles. By having the information for the slag characteristics, the study can be further investigated by predicting the slag sticking probability. Since not all the coals are made up with the same mineral components, it is required to have user defined function to simulate the slag growth. It is important to understand how much energy does the slag droplet has in order to predict its mechanistic bouncing potential after impact [7].

Table 1 highlighted the existing numerical slagging indices correlation which used the proximate analysis of the coal ash in order to calculate the slagging indices proposed in the earlier studies. The fusibility also known as softening temperature proposed by the study of back propagation of neural networks from previous works that investigate the corresponding percentage of each minerals and oxides result towards the softening temperature alteration [3,4]. Watt and Fereday viscosity model utilized the neural network study by Yin [3] as the viscosity model incorporated the softening temperature calculation into the equation. Hence by calculating the constant for each minerals composition as the neural network, the composition has the ability to alter the temperature and viscosity [8]. The viscosity model of Watt and Fereday (1969) is as follows:

$$Log\,(\mu) = \frac{m \times 10^7}{(T-150)^2} + c \tag{1}$$

where, m

$$m = (0.00835)SiO_2 + (0.00601)Al_2O_3 - 0.109 \tag{2}$$

and c,

$$c = (0.0415)SiO_2 + (0.0192)Al_2O_3 + (0.0276)Equiv.Fe_2O_3 + (0.016)CaO - 3.92 \tag{3}$$

Table 2 List of coals performance based on NSI

Coal Type	Coal Brand	Shipment No	Coal Rank	Performance based on NSI
A	Coal-A	276	Sub-bituminous	Best
A		404	Sub-bituminous	Worst
B	Coal-B	30	Sub-bituminous	Best
B		149	Sub-bituminous	Worst
B	Coal-C	584	Sub-bituminous	Best
B		345	Sub-bituminous	Worst
B	Coal-D	561	Sub-bituminous	Best
B		165	Sub-bituminous	Worst
F	Coal-Y	123	Bituminous	Best
F		106	Bituminous	Worst
Q	Coal-Z	507	Bituminous	Best
Q		77	Bituminous	Worst

Softening temperature is calculated using the neural network which take into consideration the main monomers and modifying monomers as well as minerals like CaO and MgO [1,3].

where Fe_2O_3 (equivalent) is,

$$100 = SiO_2 + Al_2O_3 + CaO + MgO + Equiv.Fe_2O_3 \quad (4)$$

In coal-fired power station, the coal received is based on shipment and it is known that the properties of these coals varies according to the shipment, despite coming from the same mines. Thus much uncertainty exists and in order to systematically categorized this coals, shipments that comes from similar coals mine are divided into two, namely good and bad coals. For the purpose of this investigation, 686 coal shipments were provided by the power station. The analysis was carried out and the list is narrowed down to smaller number using the numerical slagging indices. A total of 12 coals were selected based on its respective performance. The coals were divided into bituminous and sub-bituminous coal rank and were indicated in Table 2 as well as its status on performance. The coals were of 5 coal type namely A, B, F, Q and R with the 6 coal brands are the commonly used coals in the local coal power plant from various countries such as Indonesia, South Africa, Australia and etc.

3 Results and discussion

3.1 Coal rankings

Figure 1(a) and 1(b) show the graphical representation of the coals that have been segregated to be the best performing coals and the least performing coals. The best performing coals is indicated by the least tendency to slagging potential while the latter indicates the coals that has high tendency to slagging potential based on nominal benchmark highlighted in Table 1. Observation made with the successful coal has lower plotted benchmark value as shown in Figure 1(a) in comparison to Figure 1(b). Figure 1(a) shows higher value of AFT, Silica Oxide ratio, and viscosity, hence this yields smaller slagging potential. While the other parameters such as the basic to acid ratio, Iron (II) Oxide content, sulphur content increment yield higher slagging potential. The numerical slagging indices stated that normally for higher ash content and lower viscosity value, the slagging potential will increase as well [2].

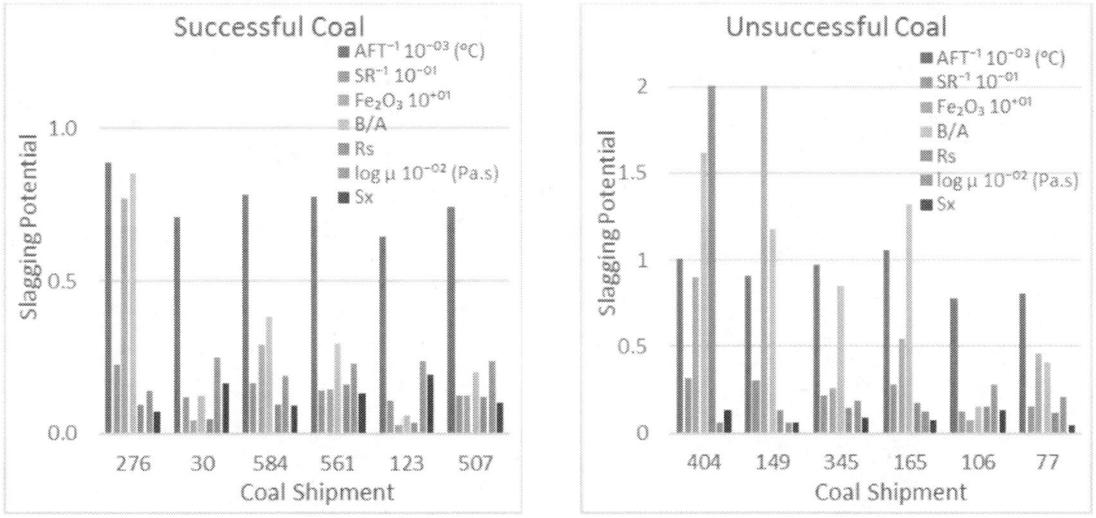

Figure 1 Comparison of slagging potential for different type of coal and shipment (a) Best performing coal (successful) and (b) Least performing coal (unsuccessful).

Figure 2 shows the comparison between the same coal brand of low slagging propensity and high slagging propensity. In this figure, the low slagging propensity coals is indicated on the left while the latter is on the right. The coals were arranged such that the best and worst performing coals of the same brand are placed side by side in order to observe the difference. Generally, high quality coals of low slagging propensity have low values for slagging indices while its counterpart of low quality coals recorded higher readings. The seven parameters are the existing indices to predict slagging propensity. The first eight coals are sub-bituminous coals with 30-Coal-B, 584-Coal-C and 561-Coal-D performed as good as the expensive bituminous coal however the overall properties of bituminous coals are better performing and the values barely exceeded 0.8.

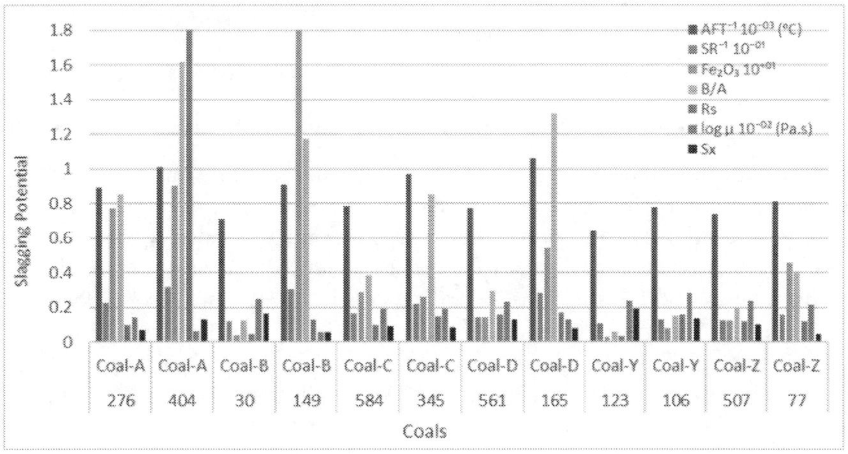

Figure 2. Comparison of the best performing coals and least performance coals of the same brand and rank.

 DOI: 10.1051/matecconf/201710905003

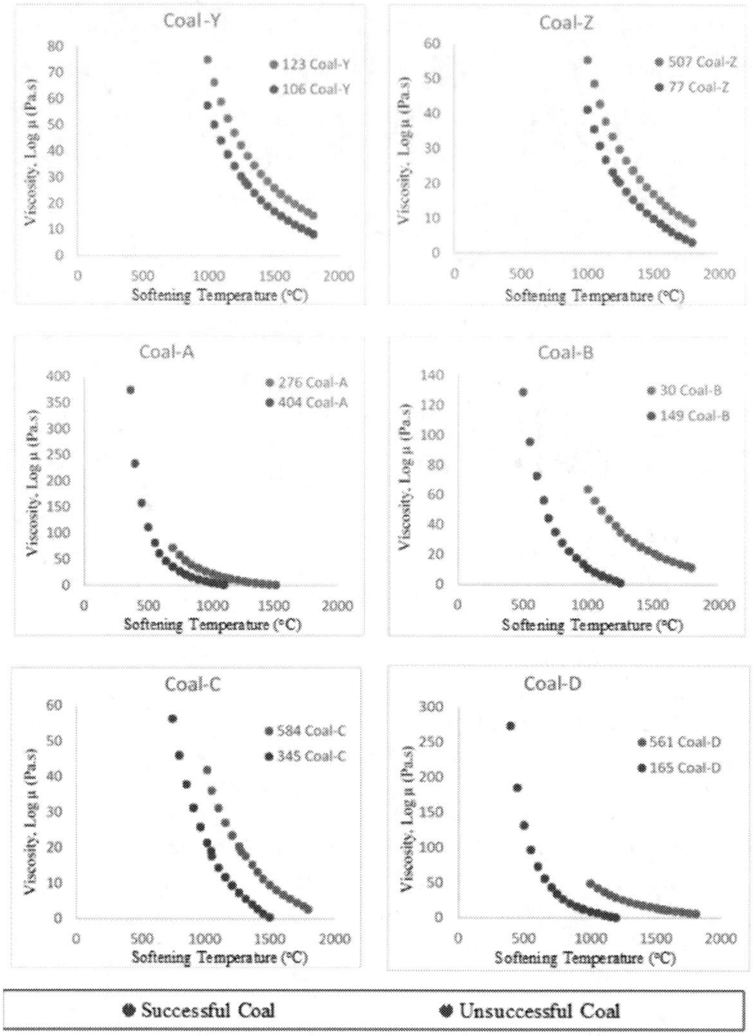

Figure 3 The comparison of viscosities and its relation to temperature for 6 different coal brands

Figure 3 shows the distribution of viscosity of the slag relationship with softening temperature. Higher softening temperature with high viscosity of slag showed low potential of slagging. Coal-A has high slagging potential as the temperature distribution at viscosity 50 Pa.s is in the region of 650°C to 800°C. Bituminous coals such as Coal-Y and Coal-Z have higher softening temperature with smaller viscosity such as 1300°C to 1500°C for Coal-Y and 1150°C to 1350°C for Coal-Z with 25 Pa.s viscosity. However, high softening temperature are required to maintain the 15 to 25 Pa.s viscosity to ensure slag is able to flow in the furnace else fluxing is required [9]. Coal-C performed fairly well as a sub-bituminous coal with the 950°C to 1250°C for 25 Pa.s viscosity for both good performing and low performing coals.

4 Conclusions

Based on the investigation carried out on bituminous and sub-bituminous coals, the mineral composition of coal ash was subjected to slagging indices, sulphur content, B/A ratio, fusibility, silica oxide ratio, numerical slagging indices, softening temperature and viscosity at the corresponding temperature. Low softening temperature of coal indicates high slagging potential. The temperature can also be integrated into the Watt and Fereday viscosity model to predict coal slag potential. In order for the slag in the molten state not to accumulate, the nominal viscosity must be in the range of 15 to 25 Pa.s. The coal ash that is not susceptible to high temperature will have reduction in boiler heat transfer efficiency as the combustion has to be tuned to have acceptable temperature distribution. Softening temperature depends on the mineral composition of the coal ash at which an increase of Silica Oxide, SiO_2 will increase the softening temperature and Iron (II) Oxide. This is also known to decrease the softening temperature as it alters the valence state of the mineral component. By understanding the slagging propensity, the coal utilization can be fine-tuned to produce enough temperature in order not to compromise the heat transfer efficiency of the furnace at the same time avoid the potential of slagging.

Acknowledgments

The authors would like to acknowledge the financial assistance received from the Ministry of Education Malaysia (MOE) under the Fundamental Research Grant Scheme (20150210FRGS).

References

1. M. U. Degereji, D. B. Ingham, L. Ma, M. Pourkashanian, A. Williams, Prediction of ash slagging propensity in a pulverized coal combustion furnace, Fuel, **101**, 171-178 (2012).
2. M. U. Degereji, D. B. Ingham, L. Ma, M. Pourkashanian, A. Williams, Numerical assessment of coals/blends slagging potential in pulverized coal boilers, Fuel, **102**, 345-353 (2012).
3. C. Yin, Z. Luo, M. Ni, & K. Cen, Predicting coal ash fusion temperature with a backpropagation neutral network model, Fuel, **15**, 1777–82 (1998).
4. A. Lawrence, R. Kumar, K. Nandakumar, K. Narayanan, A novel tool for assessing slagging propensity of coals in PF boiler, Fuel 87: 946–50 (2008).
5. J. D. Watt, F. Fereday, The flow properties of slag formed from the ashes of British coals: Part 1: Viscosity of homogenous liquid slag in relation to slag composition, Fuel, 42, 99-103 (1969).
6. A. R. McLennen, G. W. Bryant, C. W. Bailey, B. R. Stanmore, T. F. Wall. Index for iron based slagging for pulverized coal firing in oxidizing and reducing conditions, Fuel,**14**, 349–354 (2000).
7. S. Balakrishnan, R. Nagarajan, K. Karthick, Mechanistic modelling, numerical simulation and validation of slag-layer growth in a coal-fired boiler, Energy, **81**, 462-470 (2015).
8. B. Liu, Q. He, Z. Jiang, R. Xu, B. Hu, Relationship between coal ash composition and ash fusion temperatures, Fuel, **105,** 293-300 (2013).
9. P. Wang, M. Massoudi, Slag behaviour in gasifiers. Part I: Influence of coal properties and gasification conditions, Energies, **6**, 784-806 (2013).

CXCL-8 Regulates Head and Neck Carcinoma Progression through NOD Signalling Pathway

Leong-Perng Chan [1,2], Cheng Liu [3,4], Ya-Ping Tseng [5], Pin-Ju Chen [6], Da-Long Cheng [7], Wan-Ting Chen [6] and Chia-Hua Liang [6,a]

[1]*Institute of Clinical Medicine, Kaohsiung Medical University, Kaohsiung, Taiwan*
[2]*Department of Otolaryngology-Head and Neck Surgery, Kaohsiung Medical University Hospital, Kaohsiung Medical University, Kaohsiung, Taiwan*
[3]*Division of Plastic Surgery & HBOT Center, Chi Mei Medical Center, Tainan, Taiwan*
[4]*Department of Electrical Engineering, Southern Taiwan University of Science & Technology*
[5]*Institute of Basic Medical Sciences, National Cheng Kung University, Tainan, Taiwan*
[6]*Department of Cosmetic Science and Institute of Cosmetic Science, Chia Nan University of Pharmacy and Science, Tainan, Taiwan*
[7]*Department of Computer and Communication, Shu-Te University, Kaohsiung, Taiwan*

Abstract. Head and neck squamous cell carcinoma (HNSCC) ranks sixth among the most common cancers in the world. Interlukin-8 (CXCL-8), a major role in inflammatory response and tumor microenvironment, correlates with tumor progression, metastasis and invasion. We explored CXCL-8 promotes tumor progression in different differentiation HNSCC cells. This project would apply to development on biomarker and target in HNSCC as well as provide a basis of early diagnosis and treatment for clinical. CXCL-8, NOD1 (nucleotide-binding oligomerization domain-containing protein 1) and receptor-interacting protein kinase (RIPK2) levels were detected statistically higher in patient tissue with HNSCC than in non-cancerous matched tissue (NCMT) in the microarray and qRT-PCR study, whereas NOD2 was weakly expressed. Similar results were obtained for CXCL-8, NOD1, NOD2 and RIP2 from RT-PCR and western blotting. High CXCL-8, NOD1 and RIP2 expressions were found on HNSCC patient tissue than that of NCMT, whereas NOD2 was weakly expressed. The analytical results indicate that CXCL-8 is required in NOD 1-mediated signalling pathways in HNSCC.

1 Introduction

Head and neck squamous cell carcinoma (HNSCC) is the sixth most common cancer in the world. Oral cancer, predominantly oral SCC (OSCC), is a high-impact disease of the oral cavity. OSCC is generally detected in the late stages when the cancer has advanced, and therefore has a poor prognosis and survival rate [1]. Surgery and radiotherapy are currently the primary treatments, but they typically cause postoperative defects and functional impairments in patients owing to the location of OSCC in the head and neck [2]. Since morbidity and mortality rates associated with HNSCC have improved very little over the past 30 years, early detection or prevention of the disease is likely to be most effective. Therefore, basic research into HNSCC is increasing, focusing on the identification of specific biomarkers for the diagnosis of its nature and aggressiveness [3]. Cytokines are regulators of

[a] Corresponding author: tinna_ling@mail.cnu.edu.tw

host responses to infection, immune responses, inflammation, and trauma. Some cytokines make disease worse (pro-inflammatory), whereas others reduce inflammation and promote healing (anti-inflammatory) [4]. HNSCC is known to develop many molecular strategies to escape from efficient anti-tumor immune responses. Researchers do not yet well understand the origin of the signals and mechanisms underlying these responses. However, based on the local and systemic nature of these responses, researchers hypothesize that SCC produces cytokines with pro-inflammatory, pro-angiogenic and immune-regulatory activities possibly contributing to the pathogenesis of HNSCC [5]. Interleukin-8 (CXCL-8) plays a significant role in inflammatory response and tumor microenvironment; it correlates with tumor progression, metastasis and invasion, and could thus contribute to the pathogenesis of the disease [1]. Blocking IL-1, CXCL-8 or TNF has been highly successful in patients with rheumatoid arthritis, inflammatory bowel disease or graft-vs-host disease [6]. Previous research has indicated that CXCL-8 is a potential biomarker for HNSCC in bodily fluids such as blood and saliva [7]. However, different studies have found wide variations the CXCL-8 levels, and have not measured all tissue CXCL-8 levels. The NOD pathways directly regulate cytokine release, particularly of pro-inflammatory cytokines, such as IL-1β, IL-6, CXCL-8 and other immune response regulating cytokines such as IL-10 or TNF-α [8]. NLRs are expressed in various cell types, including macrophages, neutrophils, epithelial and endothelial cells, dendritic cells, and in malignant tumors, including melanoma and HNSCC [9]. CXCL-8 has a significant role in inflammatory responses in a tumor microenvironment; it affects tumor progression, metastasis and invasion, and so contributes to the pathogenesis of the disease [1]. CXCL-8 is responsible for most of the angiogenic activity that is induced by human OSCC [10]. Recent studies have suggested that nucleotide binding oligomerization domain (NOD)-like receptors (NLRs) are key modulators in inflammatory diseases [11–12]. A previous study revealed that the expression profile of NLRs in HNSCC cells differed from that in healthy epithelial cells. Stimulation by NOD1 induced an immunological response in tumor cells that differed from that in normal epithelial cells [13]. The mechanism of the CXCL-8-induced NOD pathway from HNSCC and its regulation are poorly understood. Therefore, identifying its role in tumorigenesis can lead to novel treatment modalities for HNSCC.

2 Methods and materials

2.1 Tissue samples

All tissue samples were fresh frozen in liquid nitrogen until RNA and protein purification and microarray experiments and collected with patient consent under approval of the institutional review boards from all participating institutions. For Immunohistochemistry, HNSCC and non-cancerous matched tissue (NCMT) tissue were fixed in buffered formalin and embedded in paraffin, sliced into 3 μ m sections, and mounted on glass slides. For RNA and protein isolation, HNSCC and NCMT tissue were placed into liquid nitrogen for further biochemical analyses. Flash-frozen tissue was removed from the freezer and placed into the mortar. Tissue was ground to a fine powder and scraped into a liquid nitrogen-cooled microcentrifugetube.

2.2 Microarray

For the expression profiling of HNSCC and NSCT, total tissue RNA was extracted using Trizol reagent (Invitrogen, Carlsbad, CA, USA). An RNeasy Mini Kit (Qiagen, Hilden, Germany) was used to quantify the isolated. RNA isolated at OD 260 nm and a bioanalyzer (Agilent Technology, USA) was used to conduct a qualitative. A 0.2 g mass of total RNA was amplified using a Low-Input Quick-Amp Labeling Kit (Agilent Technologies, USA) and labeled with Cy3 (CyDye, Agilent Technologies, USA) during the in vitro transcription process. A 0.6 g mass of Cy3-labled cRNA was fragmented to a men size of approximately 50-100 nucleotides by incubation with fragmentation buffer at 60°C for 30 minutes. Correspondingly fragmented labeled cRNA was then pooled and hybridized using an Agilent

Sure Print G3 Human V2 GE 8 × 60 K Microarray (Agilent Technologies, USA) at 65°C for 17 h. After they were washed and dried by blowing with a nitrogen gun, the microarrays were scanned using an Agilent microarray scanner (Agilent Technologies, USA) at 535 nm for Cy3. Scanned images were analyzed using Feature Extraction 10.5.1.1 software (Agilent Technologies, USA), which is image analysis and normalization software. The signal and background intensity were quantified for each feature.

2.3 Quantitative real-time PCR (QRT-PCR) of CXCL-8 mRNA

QRT-PCR analysis was evaluating the mRNA expression of CXCL-8 and NOD1 in HNSCC and NCMT in human tissue to verify the results for the microarray (Welgene Biotech. Co., Taipei, Taiwan).Total RNA was extracted using a Total RNA Miniprep Purification Kit (GeneMark). qRT-PCR was conducted using specific primers. The amplification reactions were carried out in a 20 μl mixture, using SYBR Green Supermix (Bio-Rad Laboratories). All PCRs were normalized to internal control β-actin mRNA.

2.4 RT-PCR analysis

For tissue RNA isolation, the RNA was extracted as Total RNA Miniprep Purification Kit (GeneMark). For cellular RNA extraction, RNAs were isolated from vehicle control or 10 and 100 ng/ml concentrations of CXCL-8-treated cells for 72 h using the Trizol reagent (Invitrogen, Carlsbad, CA, USA). Single stranded cDNA was transcribed by priming with oligo-dT with SAMscript reverse transcriptase (GeneMark, Taipei, Taiwan). PCR amplification of the cDNA was done in a reaction mixture containing Taq polymerase (GeneMark, Taipei, Taiwan). The amplified RT-PCR products were electorphoresed on a 2% agarose gel, visualized by ethidium bromide staining and photographed under ultraviolet light. All results were normalized to the house-keeping β-actin gene. The obtained bands were visualized by image J software.

2.5 Western blot analysis

For tissue protein extraction, the frozen tissue were homogenized in lyses buffer (50 mM HEPES, pH 7.5, 150 mM NaCl, 10% glycerol, 1% Triton X-100, 1mM EDTA, 1 mM EGTA, 50 mM NaF, 1 mM sodium orthovanadate, 30 mM p-nitrophenyl phosphate, 10 mM sodium pyrophosphate, 1 mM phenylmethylsulfonyl fluoride, 10 μg/ml aprotinin, and 10 μg/ml leupeptin). Supernatants were collected and analyzed. Tissue extracts and cell lysates were prepared and subjected to western blotting with antibodies against CXCL-8, NOD1, NOD2, RIP2, CXCR1 and CXCR2, respectively, as described. Proteins were visualized with ECL reagent and their relative expression was determined by densitometry using ImageJ software program and normalized relative to β-actin.

2.6 Statistical analysis

The results are expressed as mean ± standard deviation (SD). Statistical differences were estimated by one-way analysis of variance (ANOVA) followed by Dunnett's test or the Tukey-Kramer test. A p value of 0.05 was regarded as significant. The data were analyzed and the figures potted using software (SigmaPlot Version 8.0 and SigmaStat Version 2.03, Chicago, IL).

3 Results and Discussion

3.1 Microarray, CXCL-8 and NOD signalling pathway

This investigation assessed the effect of CXCL-8 on NOD signalling. A microarray analysis was first performed to consider the expression of innate immunity genes in HNSCC and NCMT of human tissue. Specifically, a significantly increased expression (>2-fold) was observed in several genes in the inflammasome signalling pathway in HNSCC, including CXCL-8, NOD1, NOD2 and RIP2 as compared with NCMT. The factors of increase were 9.93 for CXCL-8, 3.14 for NOD1, 4.23 for NOD2 and 6.95 for RIP2 (Figure 1A). This relationship indicates that the upregulation of CXCL-8 correlates with the overexpression of NOD-mediated signalling in HNSCC. To examine the possible role of CXCL-8 and NOD as informative biomarkers of HNSCC, as indicated in the microarray study, the levels of CXCL-8, NOD1 and NOD2 in HNSCC and NCMT tissue (n = 6) were determined using qRT-PCR. As shown in Figure1B, the mean CXCL-8 and NOD1 level was detected statistically higher (3.70-foldand 2.12-fold, respectively) in patient tissue with HNSCC than in NCMT, whereas NOD2 was weakly expressed. These results confirmed the use of CXCL-8 and NOD1 as biomarkers for HNSCC detection with microarray analysis (Figure 1A). Thus, experimental data suggest that CXCL-8 and NOD1 had higher expression in HNSCC than in NCMT, whereas NOD2 was absent.

A

Gene symbol	Gene name	Signal in NCMT	Signal in HNSCC
CXCL-8	interleukin 8	173.865	1726.490
NOD1	nucleotide-binding oligomerization domain containing 1	1189	3730
NOD2	nucleotide-binding oligomerization domain containing 2	107.840	456.603
RIPK2	receptor-interacting serine-threonine kinase 2	1271.418	8839.061

B

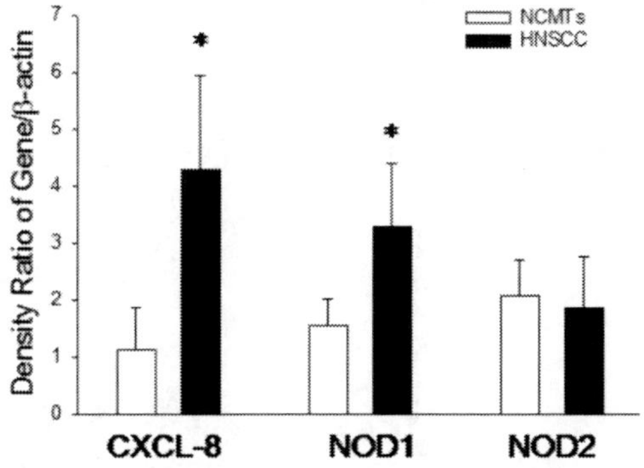

Figure 1. The expression of CXCL-8 and NOD signaling pathway in NCMT and HNSCC of human tissue. A microarray analysis to assess the expression (fold change threshold >2) of CXCL-8, NOD1, NOD2 and RIP2 in NCMT and HNSCC of human tissue. (B) The microarray data were substantially verified by qRT-PCR: CXCL-8, NOD1 and NOD2 were upregulated in HNSCC, as compared with NCMT tissue (n = 6); *p<0.05 indicates a significant difference from NCMT.

3.2 CXCL-8 and NOD1 activation in HNSCC

NOD1 induces inflammation by stimulating cellular signalling pathways through the adaptor molecule RIP2. This process in turn induces NF-κB as well as p38 and JUN amino-terminal kinases (JNK) MAP kinases. To confirm and complement the mRNA data, RT-PCR (Figure 2A) and western blotting (Figure 2B) also confirmed that the CXCL-8 and NOD pathway-related gene and protein expressions (NOD1 and RIP2) were higher in HNSCC tissue than in NCMT. Similar results were obtained for CXCL-8, NOD1, NOD2 and RIP2 from RT-PCR and western blotting. High CXCL-8,

NOD1and RIP2 expressions were found on HNSCC patient tissue than that of NCMTs, whereas NOD2 was weakly expressed. Additionally, immunohistochemical staining was performed to demonstrate the morphological localization of the CXCL-8, NOD1 and RIP2.

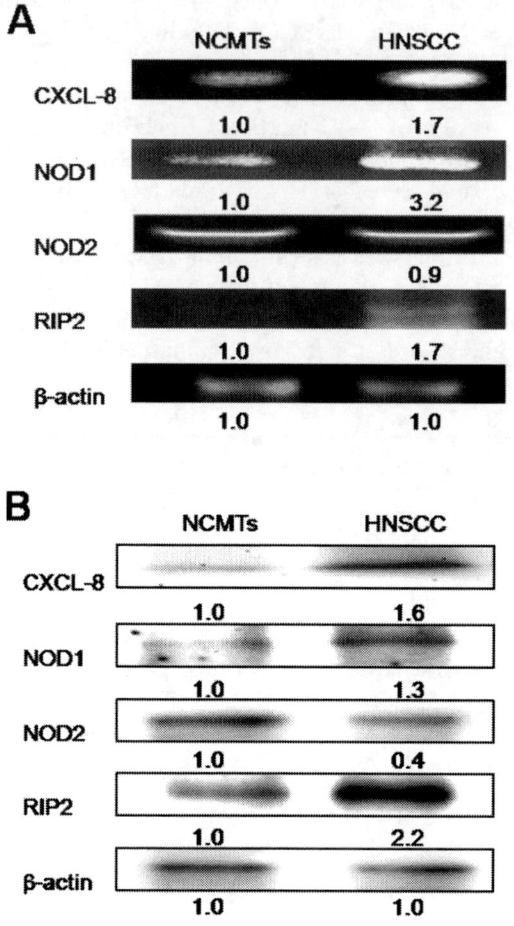

Figure 2. CXCL-8 and NOD signalling pathway in HNSCC. CXCL-8, NOD1, NOD2, RIP2 expressions in NCMT and HNSCC of human tissue were analyzed by RT-PCR (A) and western blotting (B). Quantities of β-actin were adopted as internal controls for sample loading.

4 Conclusions

Previously reports have demonstrated salivary IL-8 levels in oral lichen planus (OLP) patients are significantly decreased after dexamethasone treatment, which indicates that the levels of IL-8 could be influenced by the degree of inflammation in OLP. The present study was conducted to test the hypothesis, in the presence of chronic inflammation, irritation and carcinogenic exposure of oropharyngeal mucosa, IL-8 will be elevated leading to further activation of NOD signalling pathway contributing to a local environment which may be associated with progression of these lesions to human HNSCC. Since IL-8 play and important role in pathogenesis of HNSCC, and have been linked with increased tumor growth and metastasis, and we found that NOD pathway-mediated protein was increased by IL-8 stimulation in HNSCC cells. Our results demonstrate that IL-8 not only enhances in HNSCC tissue, but also have a stimulatory effect on the activity of NOD pathway. Further validation studies may need to be performed using a larger sample cohort for a greater statistical comparison between HNSCC and healthy control groups, which should be included to validate firmly IL-8 and

NOD1 as biomarkers for HNSCC. Therefore, if its role in tumorigenesis can be sufficiently assessed, it could open up new avenues to fine out novel treatment modalities for HNSCC.

References

1. M.A. John, Y. Li, X. Zhou, P. Denny, C.M. Ho, C. Montemagno, W. Shi, F. Qi, B. Wu, U. Sinha, R. Jordan, L. Wolinsky, N.H. Park, H. Liu, E. Abemayor, D.T. Wong. Arch Otolaryngol Head Neck Surg. **8** (2004)
2. P.J. Thomson, J. Wylie. Int J Oral Maxillofac Surg. **2** (2002)
3. M.E. Arellano-Garcia, S. Hu, J. Wang, B. Henson, H. Zhou, D. Chia, D.T. Wong. Oral Dis. **8** (2008)
4. B.B. Aggarwal , S. Shishodia , S.K. Sandur, M.K. Pandey, G. Sethi. Biochem Pharmacol. **11** (2006)
5. N.L. Rhodus, C.S. Miller, S. Myers, F. Ondrey. Cancer Detect Prev. **1** (2005)
6. S.E. Lee, J.M. Kim, S.K. Jeong, J.E. Jeon, H.J. Yoon, M.K. Jeong, S.H. Lee. Arch Dermatol Res. **10** (2010)
7. Y.S. Hwang, S.K. Lee, K.K. Park, W.Y. Chung. Oral Oncol. **1** (2012)
8. O.S. Shin, J.B. Harris. Korean J Hematol. **2** (2011)
9. F. Qiu, A. Maniar, M.Q. Diaz, A.I. Chapoval, A.E. Medvedev. Innate Immun. **4** (2011)
10. S.E. Lee, J.M. Kim, S.K. Jeong, J.E. Jeon, H.J. Yoon, M.K. Jeong, S.H. Lee. Arch Dermatol Res. **10** (2010)
11. S.R. Punyani, R.S. Sathawane. Clin Oral Investig. **2** (2013)
12. N.L. Rhodus, V. Ho, C.S. Miller, S. Myers, F. Ondrey. Cancer Detect Prev. **1** (2005)
13. Y.S. Cheng, L. Jordan, L.M. Gorugantula, E. Schneiderman, H.S. Chen, T. Rees. J Periodontol. **7** (2014)

Nonlinear Finite Element Solution of Post-buckling Responses of FGM Panel Structure under Elevated Thermal Load and TD and TID Properties

Subrata Kumar Panda[1,a], Trupti Ranjan Mahapatra[2] and Vishesh Ranjan Kar[3]

[1]*NIT Rourkela, 769008, Odisha, India*
[2]*KIIT University, Bhubaneswar, 751024, Odisha, India*
[2]*VIT University, Vellore, 632014, Tamil Nadu, India*

Abstract. The nonlinear finite element solutions for the buckling and post-buckling responses of the functionally graded shell panel subjected to the non-uniform thermal environment have been presented in this article. The thermal fields are assumed as uniform, linear and nonlinear temperature rise across the thickness of shell panel and the properties of each constituent are considered to be temperature dependent. The effective material properties of the graded structure are evaluated using the Voigt's micromechanical rule in conjunction with power-law distribution. For the analysis purpose, a general nonlinear mathematical model of the functionally graded shell panel has been developed based on the higher order shear deformation theory and Green-Lagrange type geometrical nonlinear strains. The system governing equation of the panel structure is derived using the variational principle. Further, suitable nonlinear finite element steps have been adopted to discretize the model for the computation of the desired responses in association with the direct iterative method. The convergence and the validation behavior of the present numerical model are initially tested to demonstrate its efficacy and significance. Finally, the effects of curvature, power law index and different support conditions on the buckling and post-buckling responses of the functionally graded shell panels are investigated and discussed in details.

1 Introduction

The ability of the functionally graded materials (FGMs) to withstand very high thermal gradient have made them popular in many high performance engineering applications under extreme temperature loading conditions. During service under sever thermal load, the FGM structures buckles and geometrical nonlinearity is induced. The consideration of the temperature dependent (TD) material properties (rather merely considering temperature independent (TID) material properties alone) is very much essential for the exact analysis of their stability behaviour. The linear and nonlinear buckling and post-buckling behaviour of FGM flat/curved panel subjected to uniform and non-uniform temperature loading conditions was investigated numerically by various researchers in past by using the existing/refined shear deformation theories such as the Classical Plate Theory (CPT) [1-2], First

[a] Corresponding author: call2subrat@gmail.com, pandask@nitrkl.ac.in

Order Shear Deformation Theory (FOSDT) [3] and High Order Shear Deformation Theory (HOSDT) [5]. It is evident from the review of literature that, the FEM based numerical studies related to the geometrically nonlinear buckling and the post-buckling behavior of FG spherical panels is very limited [6-7]. Also, studies using the CPT [8-9] /FOSDT [10-12] mid-plane kinematics are more in number in comparison to the HOSDT [13-15]. Researchers considered the through the thickness temperature variation to be uniform, linear and nonlinear and incorporated TD and TID material properties of the FG structures. However, in all the aforementioned research to represent the geometrical distortion the von-Karman type of nonlinear kinematics has been adopted in contrast to the Green-Lagrange type that accounts for the deformation behaviour of material continuum more accurately [16].

The motivation of the present research is to study the buckling and the post-bucking behavior of FG spherical shell panel by considering Green-Lagrange type geometrical nonlinearity and the HOSDT mid-plane kinematics. Moreover, the FG shell panel properties are evaluated using Voigt's rule of mixture in conjunction with the power-law of distribution by taking both the TD and TID material properties. The governing equation of FG panel is derived using the variational principle and discretised using a nine noded isoparametric Lagrangian element with nine degrees of freedom per node. A direct iterative method is employed to obtain the desired responses.

2 Nonlinear finite element formulations

Figure 1 shows a typical FG spherical shell panel with a rectangular base with length 'a', width 'b' and uniform thickness 'h', considered for the present analysis. The curvature at the mid surface (at z = 0) of the shell panel is defined as 'R'. The displacement field based on the HOSDT mid-plane kinematics as in [17]:

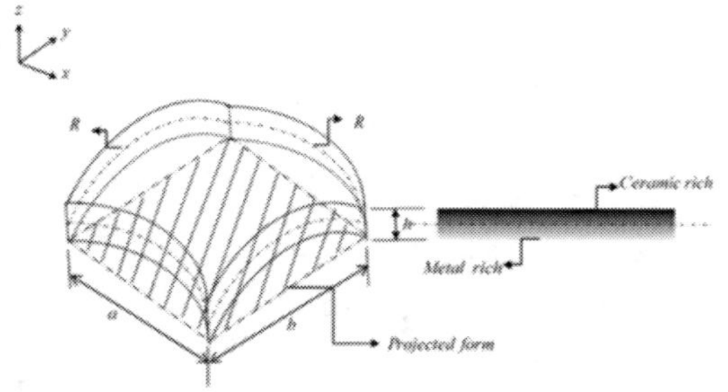

Figure 1. Geometry and dimension of FG spherical shell panel.

$$
\begin{aligned}
u(x,y,z,t) &= u_0(x,y,t) + z\theta_x(x,y,t) + z^2 u_0^*(x,y,t) + z^3\theta_x^*(x,y,t) \\
v(x,y,z,t) &= v_0(x,y,t) + z\theta_y(x,y,t) + z^2 v_0^*(x,y,t) + z^3\theta_y^*(x,y,t) \\
w(x,y,z,t) &= w_0(x,y,t)
\end{aligned}
\tag{1}
$$

where, (u_0, v_0, w_0) represent the corresponding displacements of the points on the mid-plane along (x, y, z) coordinates, respectively. θ_y and θ_x are the rotations of normal to the mid-plane about the x and y axis, respectively and $(u_0^*, v_0^*, \theta_x^*, \theta_y^*)$ are the higher order terms of the Taylor series expansion defined at the mid-plane those account for the parabolic distribution of shear stress.

The small strain and large deformation behavior of the FG shell panel is expressed by considering Green-Lagrange type nonlinear strain-displacement field as in [18]

$$\{\varepsilon\} = \{\varepsilon_l\} + \{\varepsilon_{nl}\} \tag{2}$$

where, $\{\varepsilon_l\}$ and $\{\varepsilon_{nl}\}$ are the linear and the nonlinear strain tensors, respectively.

The temperature dependent material properties of the graded structure have been obtained using the equivalent steps as in the reference [19]. Now, the effective elastic and the thermal properties of the FGM are obtained via Voigt's micromechanics model using the following equation [15]:

$$P(T,z)=\{P_c(T,z)-P_m(T,z)\}\left(\frac{z}{h}+\frac{1}{2}\right)^n+P_m(T,z) \tag{3}$$

where, subscript 'm' and 'c' represent the ceramic and the metal constituents, respectively and P_0, P_{-1}, P_1, P_2 and P_3 are the coefficients of temperature (T) and n ($0 \le n < \infty$) is the power-law index.

The temperature function is expressed in the following form for uniform temperature variation:

$$T=T_m+\Delta T \text{ or } T=T_c \tag{4}$$

where, $\Delta T=T_c-T_m$ and T_m and T_c are the temperatures at the bottom and top surfaces of the panel, respectively. The detail of the linear temperature field through the thickness condition the temperature field is expressed as can be seen in Kar and Panda [16].

The thermo-elastic stress tensor at any point within the FG spherical shell panel can be written as

$$\{\sigma\}=\{\sigma_{xx}\quad \sigma_{yy}\quad \sigma_{xy}\quad \sigma_{xz}\quad \sigma_{yz}\}^T=[C](\{\varepsilon\}-\{\varepsilon_{th}\}) \tag{5}$$

where, $[Q]$ is the reduced stiffness matrix and $\{\varepsilon_{th}\}=\{1\quad 1\quad 0\quad 0\quad 0\}^T\alpha(T,z)\Delta T$ is the thermal strain tensor at any point within the shell panel and the details of $[C]$ can be seen in [16].

The elemental displacement vector is expressed using the corresponding shape function and the nodal displacement field as

$$\{\lambda_0\}=\sum_{i=1}^{9}N_i\{\lambda_{0_i}\} \tag{6}$$

where, $\{\lambda_{0_i}\}=\left[u_{0_i}\ v_{0_i}\ w_{0_i}\ \theta_{x_i}\ \theta_{y_i}\ u_{0_i}^*\ v_{0_i}^*\ \theta_{x_i}^*\ \theta_{y_i}^*\right]^T$ is the nodal displacement vector and N_i is the shape function for the ith node, respectively. The details of the shape function can be seen in [19].

The linear and nonlinear mid-plane strain vector in terms of nodal displacement vector is written as

$$\{\bar{\varepsilon}_l\}=[B]\{\lambda_{0_i}\},\ \{\bar{\varepsilon}_{nl}\}=[A][G]\{\lambda_{0_i}\} \tag{7}$$

where, $[B]$ and $[G]$ are the product form of the differential operator matrix and the shape functions for the linear and the nonlinear strain vectors, respectively and $[A]$ is the function of displacements associated with the nonlinear strain terms. For the details of $[B]$, $[A]$ and $[G]$ matrices, [16] can be referred.

The strain energy expression can be conceded as:

$$U_\varepsilon^e=\frac{1}{2}\left(\{\lambda_{0_i}\}^T[K_l]^e\{\lambda_{0_i}\}+\{\lambda_{0_i}\}^T[K_{nl1}]^e\{\lambda_{0_i}\}+\{\lambda_{0_i}\}^T[K_{nl2}]^e\{\lambda_{0_i}\}+\{\lambda_{0_i}\}^T[K_{nl3}]^e\{\lambda_{0_i}\}\right) \tag{8}$$

The expressions for $[K_l]^e$, $[K_{nl1}]^e$ $[K_{nl2}]^e$ and $[K_{nl3}]^e$ can be seen in [18].

In the similar manner, the elemental equation of work done due to the in-plane thermal force resultant can be expressed as

$$W^e=\frac{1}{2}\{\lambda_{0_i}\}^T[K_G]^e\{\lambda_{0_i}\} \tag{9}$$

where, $[K_G]^e=\int_{-1}^{1}\int_{-1}^{1}[B_G]^T[D_G][B_G]|J|\,d\xi d\eta$ is the elemental geometric stiffness matrix.

The governing equation for the for the post-buckled FG spherical shell panel is obtained by minimizing the total energy expression. This result in

$$\delta(U_\varepsilon^e-W^e)=0 \tag{10}$$

Now, substituting Eq. (8) and (9) into Eq. (10), the final form of the equilibrium equation for the post-buckled FG spherical shell panel is conceded to the following form:

$$\left(\left([K_l]+[K_{nl1}]+[K_{nl2}]+[K_{nl3}]\right)-\gamma_{cr}\left[K_G\right]\right)\{\Delta\}=0 \tag{11}$$

where, γ_{cr} is the critical buckling load factor. $[K_l]$, $[K_{nl1}]$, $[K_{nl2}]$, $[K_{nl3}]$ and $[K_G]$ are the corresponding global matrices of $[K_l]^e$, $[K_{nl1}]^e$, $[K_{nl2}]^e$, $[K_{nl3}]^e$ and $[K_G]^e$, respectively.

3 Results and discussion

In order to compute the desired responses, a homemade computer code has been developed in MATLAB environment based on the present nonlinear FE model of the FG panel. The critical buckling and post-buckling responses are computed by considering both the TD and TID material properties. For computation, the silicon nitride (Si_3N_4) and the stainless steel (*SUS304*) are considered as the ceramic and the metal constituent at the top and at the bottom surface of the shell panel, respectively. The corresponding TD material properties are taken from the reference and details can be seen in [20].

After establishing the convergence and validation of the present model the effects of different support conditions, power-law index (n) and curvature ratio (R/a) on the post-buckling responses of FG spherical panel under different temperature fields have been investigated. The corresponding responses for FG flat panel are also presented for the sake of comparison. The different support conditions employed for the present analysis can be seen in [16].

3.1 Convergence and validation studies

First, a convergence study has been performed to note the exact mesh size at which the present model is capable of computing the linear and nonlinear post-buckling responses for the FG shell panel. The desired responses are computed for FG spherical (R/a=5, n=2, a/h=10) panel with SSSS support condition and flat (n=0.2 and 5, a/h=10) panel with CSCS support condition and presented in Figure 2 (a) and (b), respectively. The responses are obtained for different mesh sizes and amplitude ratios (Wmax/h=0, 0.5, 1 and 1.5) by considering various through the thickness temperature distribution (uniform, linear and nonlinear). It is clearly observed that the responses are converging well with mesh refinement for both the spherical and flat panels. Based on the convergence study a (5×5) mesh is used to compute the thermal post-buckling responses throughout the present study.

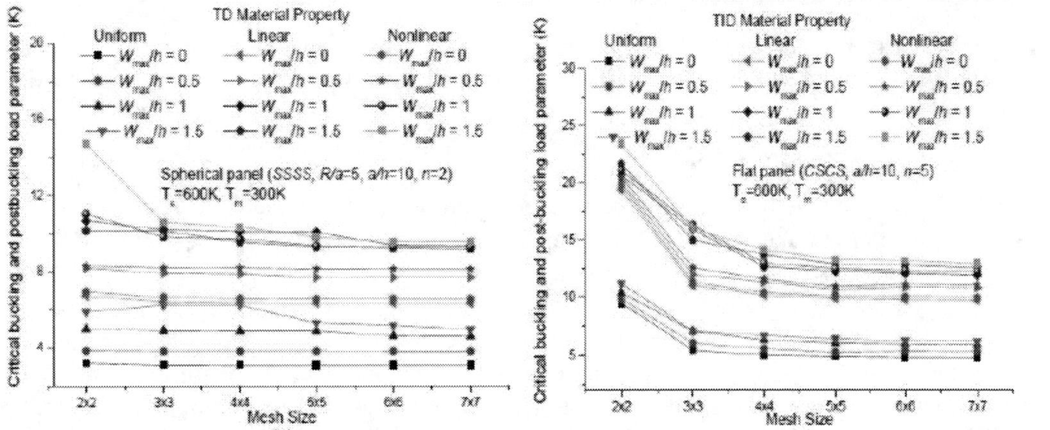

Figure 2. Convergence behavior of the thermal buckling and postbuckling responses of (a) TD-FG spherical panel and (b) TID-FG flat panel.

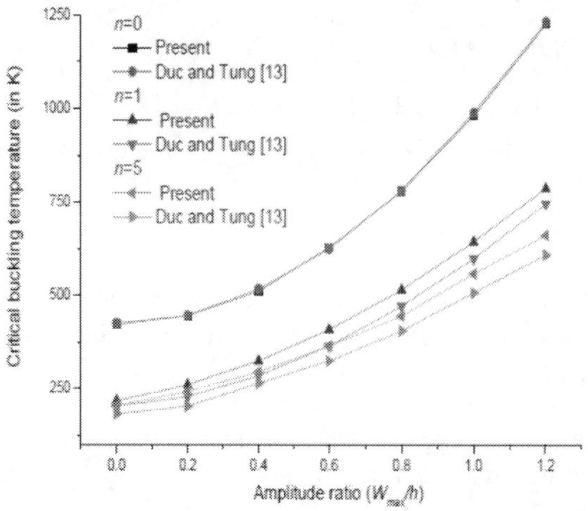

Figure 3. Comparison of the thermal post-buckling responses of simply supported FG (Al/Al2O3) flat panel.

Then the validation of the present model is offered by considering a FG (Al/Al_2O_3) flat (a/h=20 and a/b=1) panel, as the flat panels are considered to be the simplest form of the shell panel. The thermal post-buckling responses are computed under the uniform temperature rise for three different power-law indices (n = 0, 1 and 5) and presented in Figure 3 together with the available results of Duc and Tung [13]. The material and geometrical parameters are taken to be the same as in the reference [13]. It is observed that the present results are exhibiting higher critical buckling temperature values as compared to Duc and Tung [13]. It is because, the present model accounts the full geometrical nonlinearity in Green-Lagrange sense whereas, in the reference, the nonlinearity is introduced via von-Karman strain which is unrealistic to consider the exact geometrical distortion under the severe thermal load.

3.2 Additional Illustrations

In this section, some more numerical illustrations are presented to show the applicability of the present developed model to compute the post-buckling behavior of FG ($SUS304/Si_3N_4$) spherical shell panel under varying thermal loading conditions. The temperatures at the bottom and the top of the shell panel are assumed to be T_m=300K and T_c=600K, respectively and both the TD and TID material properties are considered for computation of the responses. To perform a comparative analysis, the responses are also computed for flat panel (R=∞) under identical geometry, material properties and constraint conditions.

3.2.1 Influence of temperature field

The effect of different temperature fields namely uniform, linear and nonlinear temperature distribution through the thickness on the buckling and the post-buckling responses (γ_{cr}) of the square (a/b=1) FG ($SUS304/Si_3N_4$) spherical (R/a=10) and flat panel (a/h=10, n=2) has been examined in this example. The responses are computed for different support conditions (SSSS and CSCS) and presented in Table 1. It is observed that the critical buckling and post-buckling load parameters are highest for the nonlinear temperature distribution and lowest for all uniform condition. The post-buckling responses are higher for spherical panel than the flat ones. However, the FG spherical panel with the TD properties is showing the inferior buckling and the post-buckling strength in comparison to the FG spherical panel with TID properties.

Table 1. Effect of support conditions on the critical buckling and post-buckling responses (γ_{cr}) of the FG spherical and flat panel.

Temperature field	TD/ TID	Support conditions	Flat Panel W_{max}/h				Spherical (R/a=10) W_{max}/h			
			0	0.5	1	1.5	0	0.5	1	1.5
Uniform	TID	SSSS	3.096	3.333	4.054	5.223	3.179	3.699	4.634	4.921
		SCSC	5.247	5.664	6.612	6.890	5.328	6.335	6.627	6.983
	TD	SSSS	2.728	2.942	3.565	3.788	2.807	3.283	4.150	4.471
		SCSC	4.914	5.499	7.027	7.126	5.004	6.037	7.558	7.256
Linear	TID	SSSS	6.420	6.971	8.569	8.908	6.581	7.524	9.640	10.154
		SCSC	10.903	12.148	13.493	14.373	11.051	13.172	13.703	14.792
	TD	SSSS	5.727	6.058	7.613	8.883	5.870	6.869	8.776	9.065
		SCSC	9.721	10.832	12.027	12.701	9.852	11.738	12.283	12.855
Non Linear	TID	SSSS	6.628	7.175	8.749	29.375	6.793	7.948	10.130	10.431
		SCSC	11.254	12.541	13.963	14.714	11.406	13.548	14.234	15.001
	TD	SSSS	5.923	6.417	7.866	8.224	6.070	7.115	8.275	9.364
		SCSC	10.054	11.202	14.139	13.135	10.188	12.124	12.703	13.394

3.2.2 Influence of power-law index

In this example, the effect of power-law index on the buckling and the post-buckling responses (γ_{cr}) of simply supported thick (a/h=10) square FG ($SUS304/Si_3N_4$) spherical (R/a=5) and flat panel with TD and TID properties has been investigated and presented in Figure 4 (a)-(c), respectively. It is noted that, as the FGM becomes metal rich, the post-buckling responses for both spherical and flat panel are decreasing monotonously. This is expected as the ceramic has comparatively higher stiffness than the metal. Also, the FG flat and spherical panel show highest buckling and the post-buckling strength under the nonlinear temperature field in comparison to the other temperature fields.

3.2.3 Effect of curvature ratio

It is inferred from the previous examples that the spherical panels exhibit higher buckling and post-buckling strength over the flat panel due to their spatial curvature. Therefore, the buckling and the post-buckling responses (γ_{cr}) of the thin (a/h=50) shallow FG spherical shell panel (a/b=1, n=2 and CSCS) under the effect of various curvature ratios (R/a=10, 50, 100) have been investigated in this example. The post-buckling responses are computed for uniform, linear and nonlinear temperature fields and presented in Figure 5 (a)-(c), respectively. It is observed that the buckling and the post-buckling load parameters are decreasing as the panel tend to become flat.

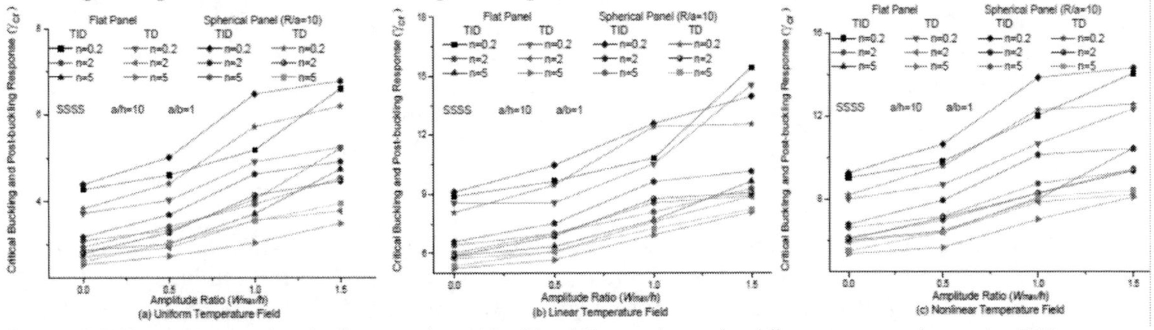

Figure 4. Effect of power-law indices on the critical buckling and post-buckling responses (cr) of FG spherical (R/a=10) and flat panel under (a) Uniform (b) Linear and (c) Nonlinear temperature field.

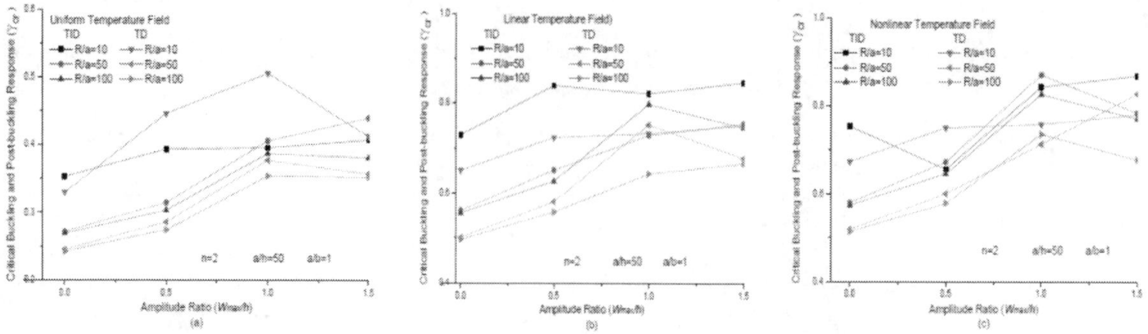

Figure 5. Effect of curvature ratio on the critical buckling and post-buckling responses (γ_{cr}) of FG spherical panel under (a) Uniform (b) Linear and (c) Nonlinear temperature field.

4 Conclusion

In this article, the buckling and post-buckling strength of FGM structure have been investigated using a new higher-order model under elevated thermal filed as well as different geometrical configurations. The effective material properties of the FGM are evaluated via the Voigt's micromechanical model in conjunction with the power-law distribution. In addition to this, both the TD and TID material properties of the FG panel are considered in the present analysis. It is observed from the present results that the TD material properties and the Green-Lagrange type geometrical nonlinearity in conjunction with the HOSDT model for the thermal stability of the FG panel structures are inevitable. It is inferred that as the power-law index increase the instability of the FGM panel is enhanced.

References

1. S. A. M. Ghannadpour, H. R. Ovesy. M. Nassirnia, *Comput. Struct.* **108(109)**, 93–99 (2012)
2. H. Huang, Q. Han, N. Feng, X. Fan, *Mech. Adv. Mater. Struc.* **18(5)**, 337–46 (2011)
3. X. Zhao, Y. Y. Lee, K. M. Liew, *Compos. Struct.* **90(2)**, 161–171 (2009)
4. H. S. Shen, *Int. J. Mech. Sci.* **49(4)**, 466–478 (2007)
5. H. Thai, S. Kim, *Compos. Struct.* **128**, 70-86 (2015)
6. K. S. Na, J. H. Kim, *Composites Part B* **35(5)**, 429–437 (2004)
7. N. D. Duc, T. Q. Quan, *Mech. Compos. Mater.* **48(4)**, 435-448 (2012)
8. H. S. Shen, *Int. J. Solids. Struct.* **41(7)**, 1961–1974 (2004)
9. T. L. Wu, K. K. Shukla, J. Huang, *Compos. Struct.* **81(1)**, 1–10 (2007)
10. J. S. Park, J. H. Kim, *J. Sound Vib.* **289(1-2)**, 77–93 (2006)
11. T. Prakash, M. K. Singha, M. Ganapathi, *Engg. Struct.* **30(1)**, 22–32 (2008)
12. K. M. Liew, J. Yang, S. Kitipornchai, *J. Appl. Mech.* **71(6)**, 839–850 (2004)
13. N. D. Duc, H. V. Tung, *Compos. Struct.* **93(11)**, 2874–2881 (2011)
14. H. S. Shen, H. Wang, *Aerosp. Sci. Technol.* **38**, 9-19 (2014)
15. Kar V R and Panda S K 2015 *Composite Structures* **129** 202–212, (2015)
16. J. N. Reddy, C. D. Chin, *J. Therm. Stresses* **21**, 593–626 (1998)
17. J. N. Reddy, *Mechanics of laminated Composite Plates and Shells*, Second Edition CRC Press (2004)
18. R. D. Cook, D. S. Malkus, M. E. Plesha, R. J. Witt, *Concepts and applications of finite element analysis*, Fourth Edition John Willy and Sons (Asia) Pvt. Ltd., Singapore (2009)
19. X. L. Huang, H. S. Shen, *Int. J. Solids Struct.* **41(9-10)**, 2403–2427 (2004)

Microstructure and Corrosion Resistance Property of a Zn-Al-Mg Alloy with Different Solidification Processes

Guang-rui Jiang[1,2,a], Ling-feng Chen[1], Hai-quan Wang[1,3] and Guang-hui Liu[1]

[1]*Shougang Research Institute of Technology, Beijing 100043, China*
[2]*Laboratory of Green Recyclable Process for Iron & steel Production Technology, Beijing 100043, China*
[3]*National Engineering Lab of Advanced Coating Technology for Metal Materials, China Iron and Steel Research Institute Group (CISRI), Beijing, 100081*

Abstract: Zn-Al-Mg alloy coating attracted much attention due to its high corrosion resistance properties, especially high anti-corrosion performance at the cut edge. As the Zn-Al-Mg alloy coating was usually produced by hot-dip galvanizing method, solidification process was considered to influence its microstructure and corrosion properties. In this work, a Zn-Al-Mg cast alloy was melted and cooled to room temperature with different solidification processes, including water quench, air cooling and furnace cooling. Microstructure of the alloy with different solidification processes was characterized by scanning electron microscopy (SEM). Result shows that the microstructure of the Zn-Al-Mg alloy are strongly influenced by solidification process. With increasing solidification rate, more Al is remained in the primary crystal. Electrochemical analysis indicates that with lowering solidification rate, the corrosion current density of the Zn-Al-Mg alloy decreases, which means higher corrosion resistance.

1 Introduction

To get high corrosion resistance of steel, several kinds of coatings were applied, such as galvanized coating [1], Galfan coating [2], Galvalume coating [3] and so on. Comparing with these coatings, Zn-Al-Mg alloy coating with much higher corrosion resistance in atmosphere and excellent corrosion resistance at cut edge attracted much attention in the last decades [4-7]. In the 80s last century, Several Zn-Al-Mg coatings composed with 6 wt.% to 11 wt.% Al and 0.5 wt.% to 3.0 wt.% Mg were applied in Japan to prolong life of buildings[8]. Due to bad welding property and bending property with high aluminium content, some Zn-Al-Mg coatings with low aluminium content was developed in European for automobile manufacture [9].

It is considered that the microstructure of alloy play important role on the corrosion resistance properties, such as Mg-Al alloys [10], Mg-Zn alloys [11] and Zn-Al alloys [12]. It is thought that the solidification process markedly effect the microstructure of alloys by casting [13-14]. However, there exist less studies on the effect of solidification process on the microstructure of the Zn-Al-Mg alloy.

This research is focus on the effect of solidification process on the microstructure of a Zn-Al-Mg alloy with low aluminium content. With different solidification processes, the electrochemical properties were studied to reveal corrosion resistance.

[a] Corresponding author: guangrui82@qq.com

2 Experimental

2.1 Material

The Zn-Al-Mg alloy was produced by casting method. Pure Zn ingot with purity of 99.995 wt.% and pure Al ingot with purity of 99.6 wt.% were melted in a graphite crucible within a pit type resistance furnace at 500 °C. Then, a master alloy composed with 11 wt.% Mg and balanced Zn was dipped into the melted alloy within high purity argon gas to avoid oxidation of Mg. The liquid Zn-Al-Mg alloy was homogenized by mechanical stirring at 450 °C. After holding 1 hour, the melt was cooled to room temperature by three different processes. Part of the melt was poured into a graphite mold chilled by water (thereafter as "water quench") and a graphite mold with same dimension mounted in the air (thereafter as "air cooling"), respectively. The remaining melt was poured into a graphite mold in the furnace (thereafter as "furnace cooling"). So, the solidification rate of water quench is the fastest and that of furnace cooling is the slowest.

The casted Zn-Al-Mg alloy ingots were cut into cubic samples by 20 mm for tests. The composition of the alloy was tested by a SPECTRO ARCOS inductively coupled plasma spectrometer, as shown in Table 1.

Table 1. Chemical composition of the Zn-Al-Mg alloy (wt.%)

Zn	Al	Mg	Bal.
97.30	1.32	1.10	0.28

2.2 Analysis methods

The microstructure of samples was explored on a S-3400N scanning electron microscope (SEM). Attached EDS was used to investigate the element composition of the samples.

Electrochemical analysis was carried out on a CHI660 electrochemical analyzer. A three electro cell was used in electrochemical experiments, with samples surface exposed 1 cm^2 as the working electrode, a platinum sheet with dimension of 30 mm x 30 mm as the counter electrode and saturated calomel electrode (SCE) as the referenced electrode. Before testing, samples were ground and polished and cleaned in ethanol. Experiments were performed in 3.5 % NaCl solutions. The polarization curve was obtained at a scanning speed of 1 mV/s from -1.25 V vs. SCE to -0.75 V vs. SCE.

3 Results

3.1 Microstructure

Microstructures are shown in Fig. 1. The chemical compositions at different location are shown in Table 2. The microstructures of samples cooled by different process are composed with a Zn crystal including some Al (position 4 in Fig. 1) and a Zn-Al-Mg ternary microstructure (position 1, 2 and 3 in Fig. 1). According to the study by Liang et al [15], the equilibrium microstructure at room temperature of the Zn-Al-Mg alloy with less than 4 wt. % Al and less than 3 wt. % Mg is composed of primary crystal of Zn and ternary eutectic crystal, which contains Zn solution phase, Al solution phase and Mg_2Zn_{11} intermetallic phase. Therefore, during the solidification, the Zn crystal with slight Al (position 4 in Fig. 1) firstly solidified and residuary Al and Mg are rejected into the inter-dendritic region. For sample cooled by water quench, the primary crystal contains about 5.9 at. % Al, as shown in Table 2. As decreasing solidification rate, Al content in the primary crystal decreases markedly. For sample cooled by furnace cooling, only 2.6 at.% Al remains in the primary crystal, as shown in Table 2.

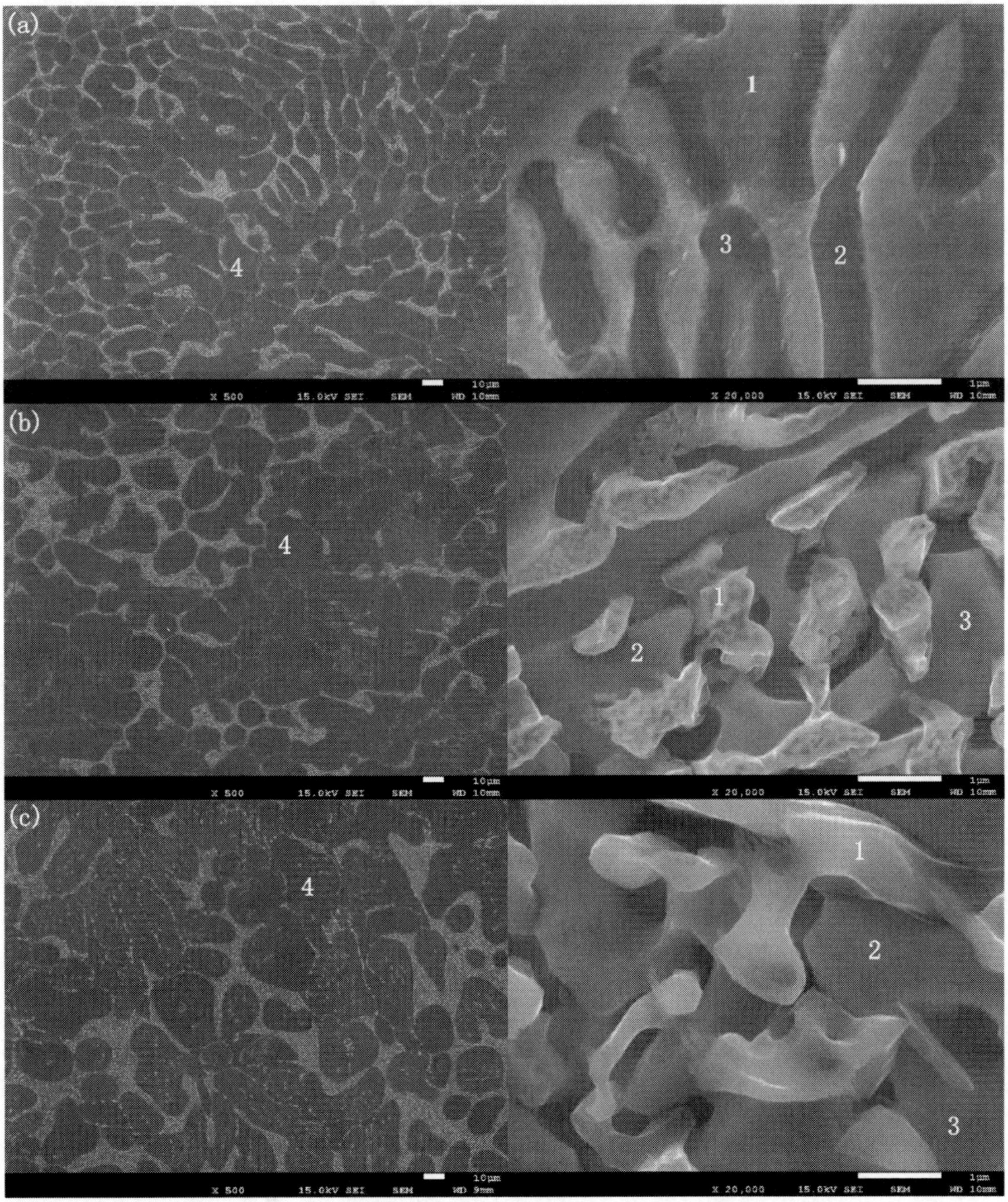

Figure 1. Microstructure of the Zn-Al-Mg alloy cooled by (a) water quench, (b) air cooling and (c) furnace cooling.

In the eutectic region, it is indicated by the EDS results that there exist three different phases with different chemical compositions, as shown in Table 2. For three different solidification processes, the atomic ratios of Mg-Zn intermetallic phase (position 3 in Table 2) in the eutectic region are similarly near to 1:2, as shown in Table 2, which indicates that the intermetallic phase is not Mg_2Zn_{11} but more likely $MgZn_2$. It should be considered to be illustrated by the non-equilibrium reaction $L \rightarrow Zn + MgZn_2$[16]. Moreover, it could be noted that the Zn content in the Al solution phase (position 1 in Table 2) in the eutectic region decreases along with decreasing solidification rate. For samples cooled by water quench and air cooling, the Zn content is 47.5 at. % and 48.1 at%, respectively. For sample cooled by furnace cooling, the Zn content in the Al solution phase decreases to about 25.5 at.%. It may due to the non-equilibrium and equilibrium monotectoid reaction (Al) → (Al) + (Zn) [17].

Table 2. Compositions of different locations in Fig. 1

Solidification	Position	Mg (at. %)	Al (at. %)	Zn (at. %)
	1	1.3	51.2	47.5
Water quench	2	5.3	7.1	87.6
	3	28.1	5.7	66.2
	4	0.0	5.9	94.1
	1	1.5	50.5	48.1
Air cooling	2	3.8	6.4	89.7
	3	32.8	2.6	64.6
	4	0.0	4.7	95.3
	1	1.2	73.3	25.5
Furnace cooling	2	0.0	3.9	96.1
	3	26.9	1.8	71.4
	4	0.0	2.6	97.4

3.2 Electrochemical properties

Fig. 2 shows Tafel curves. Electrochemical properties are shown in Table 3, which were determined by the Tafel extrapolation method [18]. The shape of Tafel curves are similar for all three processes. Corrosion current densities in anodic polarization curves increase sharply and reach to a plateau, which indicate a high dissolution behavior. In the low catholic overpotential region, the corrosion current densities increase slowly. As the value of corrosion potentials (E_{corr}) is less than-1.2 V vs. SCE, the corrosion current densities increase sharply, which implies cathodic reactions i.e. hydrogen evolution reaction and/or oxygen reduction reaction. The values of corrosion potential are similar for all samples. However, the values of corrosion current densities (I_{corr}) show much more difference. For water quench sample and air cooling sample, the values of I_{corr} are 5.8 $A{\cdot}cm^{-2}$ and 5.5 $A{\cdot}cm^{-2}$, respectively. For furnace cooling sample, the value of I_{corr} sharply decreases to 3.7 $A{\cdot}cm^{-2}$. According to standard of ASTM G 102, corrosion rate could be calculated by corrosion current density:

$$R = kI_{corr}W \quad (1)$$

where R is the corrosion rate, k is a constant value, W is the equivalent weight. Therefore, the corrosion rate is proportional to the corrosion current density for the Zn-Al-Mg alloy. So, the corrosion rate of furnace cooling sample is about sixty-three percent of that of water quench sample.

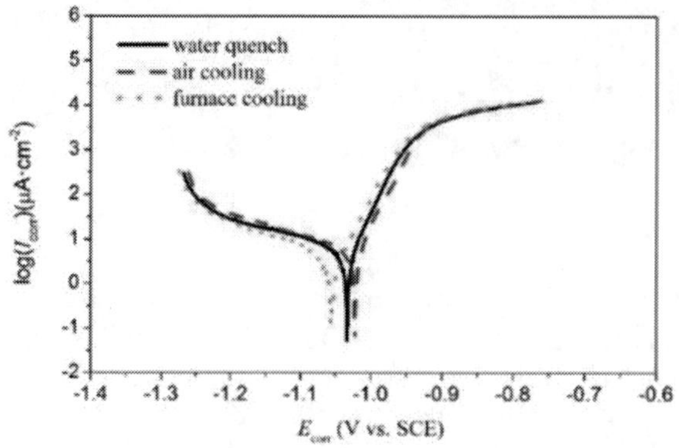

Figure 2. Tafel curves of the Zn-Al-Mg alloy

Table 3. Electrochemical properties of the Zn-Al-Mg alloy in Fig. 2

Solidification	E_{corr} *vs.* SCE (V)	I_{corr} ($\mu A \cdot cm^{-2}$)
Water quench	-1.02	5.8
Air cooling	-1.01	5.5
Furnace cooling	-1.06	3.7

4 Conclusions

A Zn-Al-Mg alloy with different solidification processes was prepared. The microstructure of the Zn-Al-Mg cast alloy is composed with a Zn crystal including slight Al and a Zn-Al-Mg ternary microstructure. As increasing solidification rate, more Al is remained in the primary crystal which is rich in Zn. Ternary eutectic microstructure is found in the Zn-Al-Mg alloy, which is compose with a Zn solution phase, a Al solution phase and a $MgZn_2$ intermetallic phase. it could be noted that the Zn content in the Al solution phase in the eutectic region decreases along with decreasing solidification rate. Tafel curves reveals that the corrosion current density of furnace cooling sample is the lowest, which implies the slowest rate of mass loss and higher corrosion resistance.

References

1. N. Pistofidis, G. Vourlias, S. Konidaris, E. Pavlidou, A. Stergiou, G. Stergioudis, Mater. Lett. **60**, 786 (2006).
2. F. Rosalbino, E. Angelini, D. Maccio, A. Saccone, S. Delfino, Electrochim. Acta **52**, 7107 (2007).
3. H. E. Townsend, J. C. Zoccola, Mater. Performance **18**, 13 (1979).
4. T. Prosek, A. Nazarov, A. Le Gac, D. Thierry, Prog. Org. Coat. **83**, 26 (2015).
5. T. Prosek, N. Larche, M. Vlot, F. Goodwin, D. Thierry, Mater. Corros. **61**, 412 (2010).
6. C. Schwerdt, M. Riemer, S. Kohler, B. Schuhmacher, M. Steinhorst, *Galvatech '04 Conference Proceedings* (Omnipress, Madison, 2004).
7. T. Shimizu, F. Yoshizaki, *Galvatech 2011 Proceedings* (Associazione Italiana di Metallurgia, Milano,2011).
8. H. Kagechika, *Proceedings of 7th International Conference on Zinc and Zinc Alloy Coated Steel Sheet* (The Iron and Steel Institute of Japan, Osaka, 2007).
9. G. Angeli, R. Brisberger, M. Bülter, L. Diez, C. Filthaut, W. Fischer, T. Koll, T. Maiwald, C. Pesci, A. Richter, J. Schulz, U. Stellmacher, N. van Landschoot, *Proceedings of 9th International Conference on Zinc and Zinc Alloy Coated Steel Sheet & 2nd Asia-Pacific Galvanizing Conference* (Metallurgical Industry Press, Beijing, 2013).
10. R. Ambat, N. N. Aung, W. Zhou, Corros. Sci., **42**, 1433 (2000).
11. E. Zhang, D. Yin, L. Xu, L. Yang, K. Yang, Mater. Sci. Eng. C **29**, 987 (2009).
12. W. R. Osório, C. M. Freire, A.Garcia, J. Alloys Compd. **397**, 179 (2005).
13. S. G. Shabestari, H. Moemeni, J. Mater. Process. Technol. **153**, 193(2004).
14. K. Honda, W. Yamada, K. Ushioda, Mater. Trans. **49**, 1395(2008).
15. H. Liang, S. –L. Chen, Y. A. Chang, Metall. Mater. Trans. A **28**, 1725 (1997).
16. T. Prosek, A. Nazarov, U. Bexell, D. Thierry, J. Serak, Corros. Sci. **50**, 2216 (2008).
17. ASM International staff, *ASM Handbook: Volume 3: Alloy Phase Diagrams* (ASM International, 1992).
18. E. McCafferty, Corro. Sci. **47**, 3202 (2005).

Design and Optimization of Die Preforming Process for Long Last-stage Blade of Nuclear Power

He Xiaomao[1,a], Jiang Peng[1], Lin Jingtang[2] and Huang Jianning[1]

[1] Beijing Mechanical and Electrical Institute of Technology, Beijing, China
[2] Tian Qian Heavy Industry Company, Ltd, Mianzhu, SiChuan, China

Abstract: The long last-stage blade is a key component of the steam turbine of nuclear conventional island. The die preforming process for a new technology that provides billets for near-net-shape roll-forging process was designed, the effects of the forging temperature, friction coefficient, flash land's height and die's outer fillet radius on the die forging force and forging energy were studied by using the orthogonal experiment method, the primary and secondary order of the four factors were analysed by using range analysis method, and the optimal combination of the factors was obtained. By means of numerical simulation and physical experiment, the die preforming process that can provide qualified billets for the subsequent roll-forging process was verified, and the PZS1120f electric screw press can meet the requirements of the die preforming process.

1 Introduction

The long last-stage blade is the key component of the steam turbine of the nuclear power conventional island. The larger the exhausting area of the last-stage blade of the LP cylinder is, the higher the efficiency of the nuclear power unit is and the better the economy is [1]. Harbin Steam Turbine Factory had successfully developed a 72-inch (1829mm) half-speed nuclear power turbine blade [2], Shanghai Electric had also developed a 67-inch (1710mm) nuclear power blade [3], the development of these long blades effectively improves the efficiency of nuclear power unit, but it's also a challenge to manufacture these long last-stage blades. At present, the technology of preforming and overall forging was mainly adopted by the blade manufacturers [4, 5]. The radial forging and the open-die forging was adopted respectively by the foreign and domestic blade manufacturers as blanking process, and then, the overall die forging process was adopted to form the blades. Due to the large area of these blades, the capability of the blade manufacturers' forging equipment cannot meet to the required forging force of producing the blades and the forged blades had got some defects, such as underpressing, underfill and overweight. The die preforming process that provides billets for a new near-net forming process for nuclear last-stage blades was proposed, the new process, which uses the roll-forging process to form the blade body of the long blades, the die forging to form the blade root, crown and damper, can effectively reduce the required forging force. In this paper, the die preforming technology of long last-stage blade was designed, the process parameters were optimized by the

[a] Corresponding Author: hexiaomaos@163.com

orthogonal experiment method and numerical simulation, and the feasibility of the process was verified by physical experiments.

2 Optimization of process parameters

The die preforming process was carried out in the PZS1120f type of electric screw press, the nominal tonnage of the press is 250,000 kN. According to the nominal tonnage, the maximal projection area of the forging billet was calculated. The forging force is not only related to the material and forging's projection area, but also related to the strain rate, temperature, friction coefficient and die structure (height of the flash land and the outer fillet radius of dies). Because the strain rate is related to the forming speeding of the press, its value is determined when the forging press is chosen. The forging temperature, the friction coefficient, the height of the flash land and the outer fillet radius were chosen as the process parameters to be optimized and the forging force and forging energy were chosen as optimization objectives.

2.1 Die preformed forging and dies

3D model of a preformed forging was shown in Figure 1, the forging model was calculated based on the press nominal tonnage and the forging drawing of the 72-inch blade, the calculation process wasn't repeated here, it will be discussed in detail in the future paper. The die's structure was shown in Figure 2, it didn't contain the groove of the flash, and the structure was quite simple, only the height of the flash land h and the outer fillet radius r were considered as parameters.

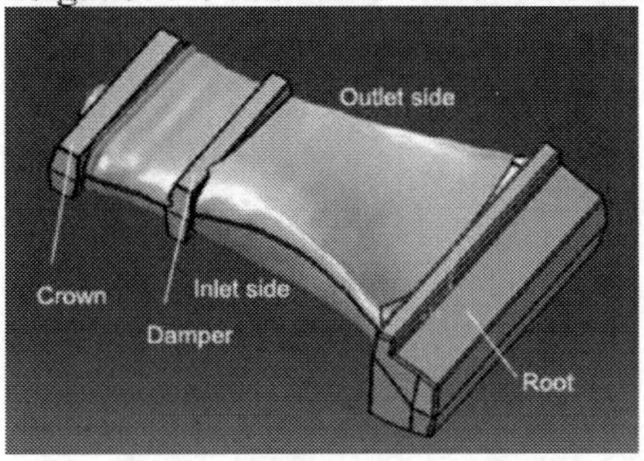

Figure 1. 3D model of die preformed forging

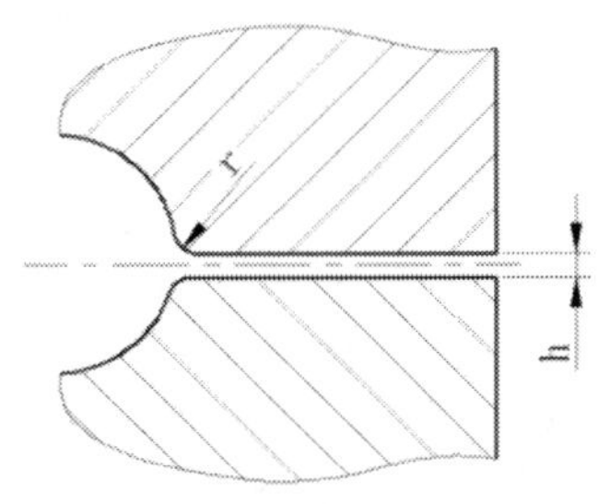

Figure 2. Die Structures

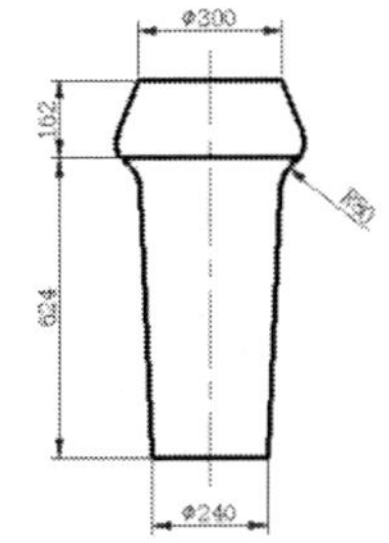

Figure 3. Billet of the die preforming

2.2 Levels and factors table of orthogonal experiment

Orthogonal experiment method can be used to study the impact of multiple factors on the optimization objective by less number of trials, to obtain the best combination of factors for the optimal objective value. The forging force and the forging energy of the die preforming are related to the forging temperature A (T), the friction coefficient B (μ), the height of flash land C (h) and the outer fillet radius D (r). The four parameters were adopted as the orthogonal experiment factors, each factor took three levels, and the designed factors and levels table were shown in Table 1.

Table 1. Factors and levels table of the orthogonal experiment.

Factor / Level	A Forging temperature T/°C	B Friction coefficient μ	C height of flash land h/mm	D outer fillet radius r/mm
1	1180	0.2	6	2
2	1150	0.3	8	3
3	1120	0.4	10	4

2.3 Parameter settings of numerical simulation

The billets used in the die preforming were produced in the semi-open heading process in ref. [6]. The shapes and dimensions were shown in Fig 3. The material is 1Cr12Ni3Mo2VN, and the constitutive relation of the material was from ref. [7], in the Arrhenius form and the hyperbolic sinusoidal was used.

$$\dot{\varepsilon} = 2.63435\times 10^{16}[\sinh(0.009236\sigma)]^{6.027}\exp\left(-\frac{453160}{8.3147T}\right) \tag{1}$$

The die was modelled as a rigid model with a preheat temperature of 200 ° C and a forming speed of 400 mm / s. The forging temperatures and the friction coefficients were set according to Table 1.

2.4 Results of orthogonal experiment

Orthogonal experiments were performed using the 3-level and 4-factor table L_9 (3^4) [8], without considering the interaction between the factors, the orthogonal experiments of simulation arrangement and the results were shown in Table 2.

Table 2. Orthogonal experiments of simulation arrangement and the results

NO.	Factor				Result	
	A Forging temperature	B Friction coefficient	C Flash land height	D Outer fillet radius	Forging Force F (kN)	Forging Energy E (kJ)
1	1	1	3	2	343000	5830
2	1	2	1	1	462000	6200
3	1	3	2	3	449000	6440
4	2	1	2	1	398000	6420
5	2	2	3	3	364000	6510
6	2	3	1	2	631000	7370
7	3	1	1	3	522000	7270
8	3	2	2	2	465000	7370
9	3	3	3	1	427000	7440

2.5 Range analysis

The value of K_{jm} was the sum of the result of the m factor at j level, and the means were represented by k_{jm}, the value of k_{jm} can reflect the optimal level. R_m was the range of the mth factor, the value of R_m reflected the fluctuation range of the optimal objective when the m factor fluctuated, the bigger the R_m was, the greater the influence of the m factor on the objective was. The primary and secondary order of the factors can be judged according to the range. The results of the range analysis of the forging force F and the forging energy E were shown in Table 3.

Table 3. Range analysis of the forging force F and the forging energy E

Result		Factor			
		A	B	C	D
Forging Force F (kN)	K_{1m}	12540000	1263000	1615000	1287000
	K_{2m}	13930000	1291000	1312000	1439000
	K_{3m}	14140000	1507000	1134000	1335000
	k_{1m}	4180000	421000	538333.3	429000
	k_{2m}	464333.3	430333.3	437333.3	479666.7
	k_{3m}	471333.3	502333.3	378000	445000
	R_m	53333.3	81333.3	160333.3	50666.7
Order		C>B>A>D			
Forging Energy E (kJ)	K_{1m}	18470	19520	20840	20060
	K_{2m}	20300	20080	20230	20570
	K_{3m}	22080	21250	19780	20220
	k_{1m}	6156.7	6506.7	6946.7	6686.7
	k_{2m}	6766.7	6693.3	6743.3	6856.7
	k_{3m}	7360	7083.3	6593.3	6740
	R_m	1203.3	576.7	353.3	170
Order		A>B>C>D			

DOI: 10.1051/matecconf/201710901005

According the range of forging force in Table 3, the order of importance that the factors influences the forging force were the height of flash land C, the coefficient of friction B, the forging temperature A, the outer fillet radius D. According to the range analysis, the optimal combination to obtain the minimum forging force was (C3B1A1D1). The relationships among the forging force and the factors were shown in Figure 4, forging force decreased with increasing temperature, with decreasing of friction coefficient, and with increasing of the flash land's height; the relationship between forging force and the outer fillet radius didn't show significant trend of increasing or decreasing, but the range was the smallest among the four factors.

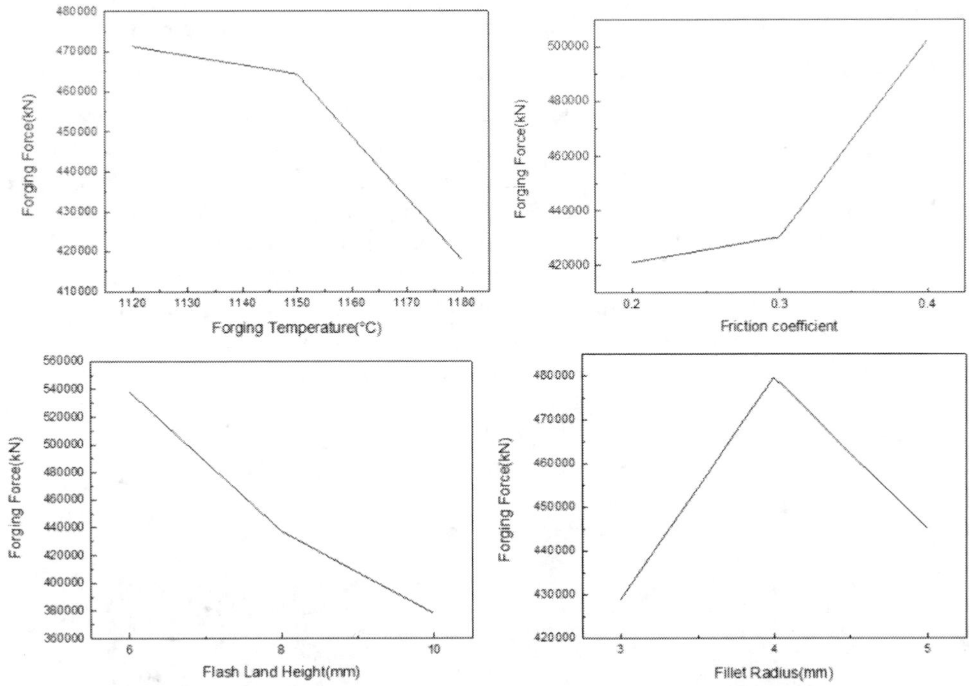

Figure 4. Relationships among the forging force and factors

According the range of forging force in Table 3, the order of importance that the factors influences the forging energy were forging temperature A, friction coefficient B, height of flash land C, outer fillet radius D. Except the forging temperature can cause some fluctuations of the range of forging energy, the effects of other factors on the forging energy's fluctuation were quite small, because the heating temperature made the transition of crystal atoms more easily, macroscopically indicated that the metal was more easily deformed, and the temperature increasing weakened the third-phase particles on pinning the dislocation movement, the required deformation energy was smaller; And the rest factors didn't have effects on the atomic migration and dislocation movement, and the macro behavior was that the forging energy didn't fluctuate obviously.

3 Numerical simulation analyses

According to the optimizing result of the orthogonal experiment, the parameters were: forging temperature was 1180 ℃, friction coefficient was 0.2, flash land's height was 10mm, and outer fillet radius was 3mm, and the other parameters were same as orthogonal experiment. The simulation results were shown as follows:

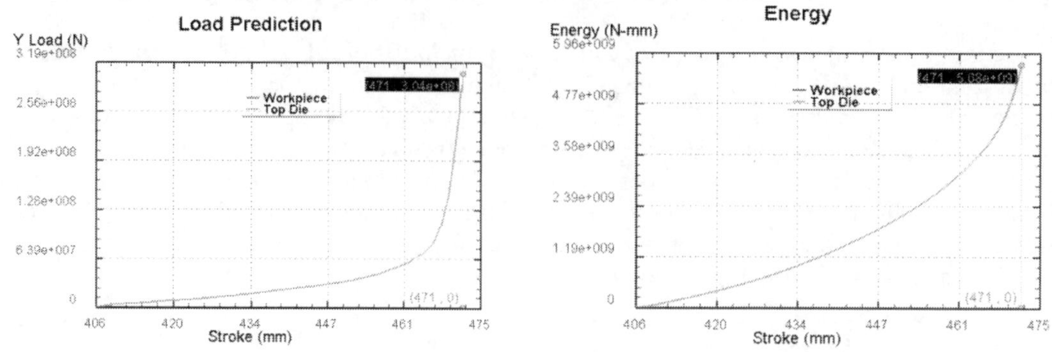

Figure 5. Relationship between stroke and forging force and energy

As shown in figure 5, the predicated forging force was 304000 kN and the forging energy was 5680 kJ by using the optimal scheme. Compared with the result of orthogonal experiment, it was found that the value simulated by the optimal factor and level combination was the smallest one.

Even if the optimal scheme was utilized, the forging force of 304,000 kN was great than the nominal tonnage 250,000 kN of PZS1120f electric screw press. In the live production, there would be some issues, such as the upper and lower die cannot be clamped, the forging would be underpressing and underfill. However, in order to ensure that forged billet has enough width and a certain length, to minimize the deformation along the width direction during the roll forging step, a few underpressing in the height direction would be allowed. The situation of die cavity filling while the forging force was 207000kN was shown in Figure 6, except the crown part of the outlet side, the rest of the billet was almost filled, and the value of underpressing of the dies was 2.7mm, this underpressing value would not have too much impact on the roll-forging of the blade.

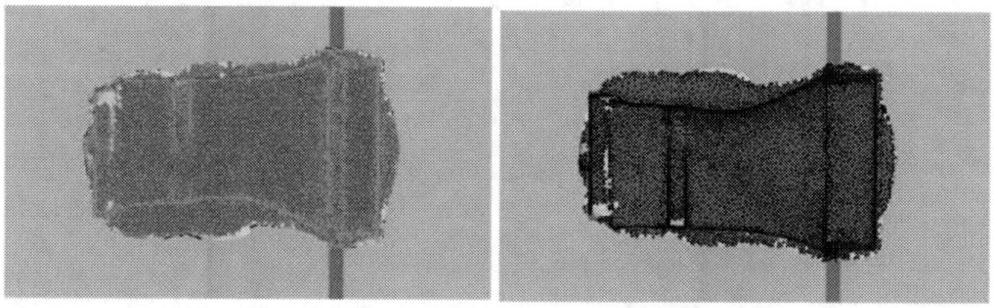

Figure 6. Situation of die cavity filling while the forging force was 207000kN

The reason that cause the outlet side of the crown was not filled was the billet deflected to the inlet side at the beginning of the forging (as shown in Fig. 7), resulting in a shortage of material on the outlet side and the flash at inlet side was too large (Figure 6).

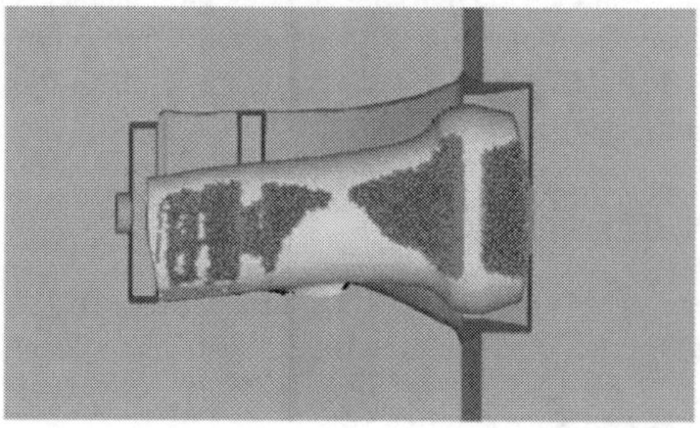

Figure 7. Deflection of the billet

DOI: 10.1051/matecconf/201710901005

4 Die preforming experiment

Figure 8a were the upper and lower die used for the die preforming experiment, and Figure 8b was the heading billet, which was forged by the heading process. Heating temperature was 1180 ℃, die preheating temperature was 200 ℃, lubricated by graphite emulsion lubricant, and two billets were forged in this experiment.

The working force was set to the nominal tonnage of PZS1120f electric screw press, during the forging process, the first blow was to locate the billet's position and to remove the scale, and then, the billet was forged by the next two blows. The final forging forces were measured to be 223460 kN and 213690 kN respectively. The die preforming forgings were shown in Figure 9, two forgings had varying degrees of underpressing and underfill, the underfill of the crown part at the outlet side was more serious, and the flash at the inlet side was much large. With numerical simulation results, it can be determined that the reason of these defects was the blank's deflection to the inlet side; some measures should be taken to prevent the blank deflection in the live production.

The flash land's heights of the two forgings were measured to be 12.9mm and 12.8mm, the values of underpressing were 2.9mm and 2.8mm, because the forging force had reached the maximum tonnage of the press, it's hard to clamping the upper and lower die by increasing the number of blows, and it would damage the press and dies. These defects can only be solved by the subsequent roll-forging process.

a) Upper and lower dies b) heading billet

Figure 8. Dies and billets of the die preforming experiment

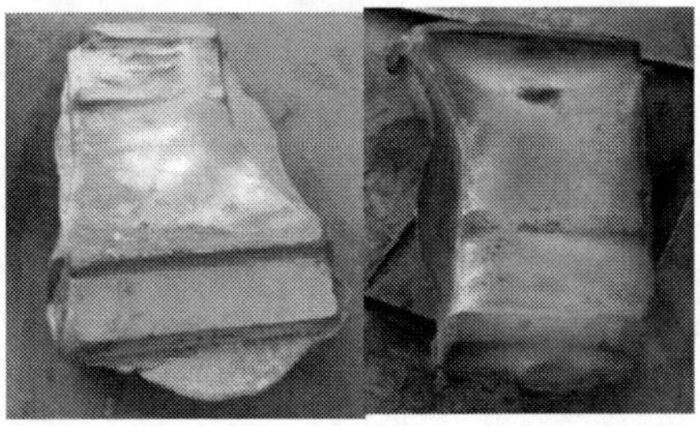

Figure 9. Die preforming forgings

The feasibility of the die preforming process was verified by the experiment, and the PZS1120f electric screw press can basically meet the requirements of die preforming process, and the qualified billets can be provided for the subsequent near-net-shape roll-forging process.

5 Conclusions

Through the orthogonal experiment optimization, numerical simulation and physical experiment, the following conclusions can be drawn:

1) Among the four factors: Forging temperature A, friction coefficient B, flash land's height C and the outer fillet radius D, the principle factor that impacts on forging force was the flash land's height, and the forging temperature had some effects on the forging energy, the optimal factor and level combination was C3A1B1D1.

2) The qualified billets formed by die preforming process can be provided for the subsequent near-net-shape roll-forging process and the PZS1120f electric screw press basically meets the requirements of the process.

3) The billet was deflected during the forging process, resulting in the underfill of the crown part at the outlet side; some measurements should be taken to prevent the deflection in the live production.

References

1. T. Zhou, M. Zhang, L. Zhang, K. Ran, China Electric Power(Technology Edition), **2**, 43, (2012)
2. Q. H. Zhang, Y. F. LI, J. W. Guang, Symp. Turb. Prof. Comm. CSPE, 133, (2012)
3. W. Lu, Z. Y. Peng , Y. Zhou, K. Cheng, East China Electric Power, **38**, 1771, (2010)
4. X. J. Li, Z. F. Huang, W. C. Chen, R. J. Qin, Power Equipment, **24**, 150, (2010)
5. J. Zhong, C. J. Hu, C. Guo, Forging & Stamping Technology, **33**, 1, (2008)
6. X. M. He, P. Jiang, F. W. Li, J. N. Huang, Forging & MetalForming, **13**, 49, (2016)
7. X. M. He, P. Jiang, J. T. Lin, Y. Yang, Journal of Plasticity engineering, **4**, 96, (2016)
8. Y. B. Qu. *Experimental Design and Data Processing* (China Univ. Sci. & Tech. Press, 2008)

AUTHOR INDEX